VIDEOCONFERENCING DEMYSTIFIED

Mc GRAW-HILL
TELECOMMUNICATIONS

Videoconferencing Demystified

Making Video Services Work

Steven Shepard

McGraw-Hill
New York   Chicago   San Francisco   Lisbon
London   Madrid   Mexico City   Milan   New Delhi
San Juan   Seoul   Singapore   Sydney   Toronto

Cataloging-in-Publication Data is on file with the Library of Congress.

McGraw-Hill

A Division of The McGraw·Hill Companies

Copyright © 2002 by The McGraw-Hill Companies, Inc. All rights reserved.
Printed in the United States of America. Except as permitted under the United
States Copyright Act of 1976, no part of this publication may be reproduced or
distributed in any form or by any means, or stored in a database or retrieval system, without the prior written permission of the publisher.

1 2 3 4 5 6 7 8 9 0   DOC/DOC   0 8 7 6 5 4 3 2

P/N 0-07-140872-X
PART OF
ISBN 0-07-140085-0

The sponsoring editor for this book was Steve Chapman and the production supervisor
was Sherri Souffrance. It was set in Century Schoolbook by MacAllister Publishing
Services, LLC.

Printed and bound by R.R. Donnelley & Sons Company.

 This book is printed on recycled, acid-free paper containing a minimum of 50
percent recycled de-inked fiber.

For my family. There are some things a videoconference just can't replace.

CONTENTS

Contents

ACKNOWLEDGMENTS

I enjoy writing these books, particularly because they give me a good excuse to stay in touch with people whom I am fortunate to call friends. They push when I slow down, cheer when I flag, and gently but firmly keep me on course when I stray.

For this book, I am particularly indebted to my good friends at Proximity, Richard Parlato, Bob Kaphan, and Bob Maurer, who opened the doors of their videoconferencing facility to me, allowed me to poke into the corners of their business, and patiently helped me understand what they do. Thank you, my friends.

I am also grateful to the following people who helped in various ways by suggesting content, reading and editing manuscripts, providing research effort, and in some cases hiring me to create and deliver content to their employees using a variety of the media described in this book. These people include Cyril Berg, Rich Campbell, Phil Cashia, Bob Dean, Mark Fei, Jack Garrett, Jack Gerrish, Barbara Jorge, Gary Kessler, Phyllis Klees, Naresh Lakhanpal, Gary Martin, Mitch Moore, Rick Sanders, Kenn Sato, Kirk Shamberger, Henry Sherwood, Elvia Szymanski, Christine Troianello, Tom Vairetta, Ken Wade, and Dave Whitmore. Thank you all for putting up with me.

Steve Chapman at McGraw-Hill has served as my editor and my friend, doing both equally well. Thanks, Steve. Jessica Hornick, also at McGraw-Hill, manages to keep Steve in line and me on track. Thanks, Jessica.

As always, I reserve this last paragraph for my family. Steve, Cristina, thank you for teaching me what it means to be a good person. Bina, thanks for—everything.

FOREWORD

More than anything else, I am a teacher—an educator. I work with corporations that design, build, and maintain networks; the companies that manufacture the switches and related hardware that go into those networks; and the component manufacturers that build semiconductor and optoelectronic devices that make those switches, multiplexers, and other devices function properly. I help these companies understand the role they play not only within the smaller sphere of telecommunications, but also within the greater spheres of economics, corporate communications, human interaction, and global responsibility.

I also work with companies that benefit from global telecommunications networks, the so-called verticals or stovepipes: health care, government, education, finance, advertising, large business, small business, and residence. I spend a lot of time in classrooms and boardrooms, helping the participants understand not only how the technologies work, but how they can be used as strategic tools to make them more competitive and effective in their own markets.

As an educator in the classroom, I have the benefit of eyeballs. I can gaze into the crowd while I'm lecturing and instantly know who is getting it and who is not, who is engaged and who is marking time, who is embracing what I have to say and who is playing the all-important role of corporate cynic, the person whose body language screams, "I've seen this before and I'm here because I have to be—get on with it." As a professional educator, I have a set of tools and techniques that allow me to work effectively with all of those perspectives, but I've always believed that I can't manage what I can't see. If I can't gaze into their eyes, the "windows of the soul," I can't anticipate what participants are thinking or feeling and therefore, can't be as effective as I want to be as I attempt to change the way they think about the subject at hand.

I've worked in telecommunications for 21 years, 11 of those in some form of education or training. I have seen the "Training Preference Pendulum," illustrated in Figure 1, swing inevitably back and forth between the merits of live, instructor or facilitator-led events, and those conducted via some form of alternative medium:

Figure 1
The pendular
movement of the
training industry,
ranging from
instructor led to
alternative media

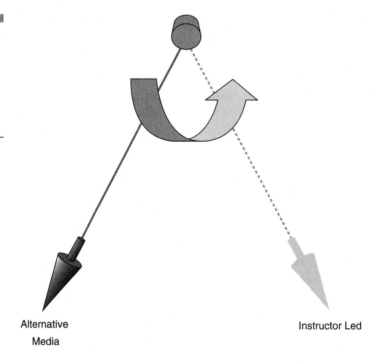

Alternative
Media

Instructor Led

videoconference, web-based conferencing, interactive CD-ROM, full-motion video, written materials, fancy variations on conference calls, and *computer-based training* (CBT). As we will discuss later in the book, all of these have their place in education, corporate meetings, and other information dissemination opportunities. Approximately every 10 years, the pendulum swings from one extreme to the other, with the attention of professional educators and facilitators swinging right along with it.

Of course, as an instructor who loves being (as my colleague Kirk Shamberger is wont to say) "the sage on the stage," I have always been biased toward the classroom environment, willing to argue vociferously that no other medium comes close to its effectiveness as an information sharing vehicle. When I started in this business in the early 1980s, I had every reason to make that argument. PCs had *barely* made their appearance, so CBT was typically nothing more than an electronic book—a digital page turner, if you will. Cost-effective video CODECs had not yet made their appearance, so videoconferencing was out of the question for all but the largest (and most daring) corporations. Even for those corporations willing to take a chance on it (telephone companies, for example, who owned the networks over which it operated), there were challenges. The video quality was so poor that attendees spent as much time pointing out the inferior nature of the medium as they did engaging in the meeting that it was attempting to facilitate. There was no Web yet, so online distance learning wasn't even a dream, and the only CD-ROMs in existence were the size of steak platters and had nothing that even came close to multimedia content. Audio conference bridging could be done, but was extremely expensive, as was full-motion video. Faced with this reality, it was easy to make the argument that the classroom remained the only truly effective method for educational pursuits and that the body-filled conference room was the only way to conduct a meeting with any success.

Over time, the makeup and capability of the network changed, and with it, my opinion about the relative efficacy of alternative delivery media. As processor power grew in concert with Moore's Law, as network access bandwidth grew and became more available with the arrival of such technologies as 56 Kbps modems, ISDN, DSL, cable modems, and broadband wireless, I began to see a glimmer of possibility for these new services.

The 1990s, often dubbed the "Age of Multimedia," saw things come full circle. Multimedia development environments, such as Quick-Time and Macromedia Director, began to produce incredibly rich applications that could take advantage of faster processors and networks. HTML and Java, both born of and nurtured by the World Wide Web, provided easy-to-use development engines that delivered value in every business, from single person operations to the largest corporations. Solid-state memory (RAM), archival memory (CDs and the like), and hardware-based memory (hard drives) plummeted in price as manufacturing techniques improved and competition for customers between incumbent and new players grew. Most important of all, those of us for whom these tools were intended and the people for whom we would use the tools began to accept the fact that perhaps these alternative delivery media—sometimes termed "telepresence technologies" by the techno-cognoscenti of the day—actually worked when used properly.

As I incorporated these tools into my work, I found that in fact they had the potential to be quite useful, in some cases—horror of horrors—more effective than the classroom. Of course, the opposite was also true. In the frenzy to embrace the pendular rise of alternative media at the crest of the 10-year cycle, many professional facilitators used them wherever they could without regard for the appropriateness of their use. I call this the "Jurassic Park Effect." Those of you who saw the movie will recall a scene at the beginning where the main characters are eating lunch (Chilean sea bass, I believe) with the park's founder, played by Richard Attenborough, and his lawyer, having just seen the dinosaurs for the first time. During the scene, it becomes clear that chaos theorist Ian Malcolm (played by Jeff Goldblum) is troubled by what he has seen. Attenborough challenges him, saying, "How can you be bothered by this? We've created dinosaurs!" Goldblum replies, "Yes, but you spent so much time and money proving that you *could*, you forgot to ask yourselves whether you *should*."[1] Just because a technology or medium is new and available does not mean that it should necessarily be used. Other factors beyond the technology must be considered because if the wrong technology-based solution is used and it fails, it will immediately be branded *bad*. Getting participants to consider its use again will be a difficult battle.

Since 1876, when A. G. Bell first emerged with his speaking telegraph machine, we have unceasingly built increasingly expensive and complex networks that envelop the globe with copper, coaxial cable, optical fiber, and radio waves. We have constructed massive central offices in every community that are stuffed with expensive hardware designed to facilitate the establishment of network connections. We have written hundreds of billions of lines of complex software code that purport to give intelligence to the world's networks. We have created the semiconductor, a device that, in a single flake of layered silicon the size of a thumbnail, replaces hardware that 30 years ago occupied a room the size of a football field. We have also created a host of applications that reside on these networks. They have names like Caller ID, Three-Way Calling, Call Forwarding, Instant Messaging, videoconferencing, and virtual reality.

---

[1] And of course, we know that they shouldn't have, because there's a sequel.

It is interesting to note that all of these applications—*without a single exception*—have one goal: to emulate, as realistically as possible, the experience of two or more people having a conversation. All the switches, multiplexers, cell towers, routers, telephones, and the millions of miles of wire and fiber provide a less-than-perfect replacement for the simple act of talking to one another.

Enter the 21st century. In early 2001, we were at the extreme end of a 10-year pendulum swing in favor of alternative media. Videoconferencing enjoyed significant uptake, and web conferencing was becoming a reality in the form of such innovative companies as PlaceWare, Centra, and Webex. Because of ISDN, DSL, and cable modems, telecommuters, *Small-Office/Home Office* (SOHO) workers, and road warriors were experimenting with personal videoconferencing. Multimedia-rich, computer-based training applications were flying off the shelves, teaching everything from telecommunications network basics to language skills.

Those of us in the business knew that the mania of popularity that these media were experiencing would be short lived, just as it had been in previous cycles. They would flare brightly for a couple of years, and then settle down to a dull glow as the pendulum swung back toward the classroom.

Then, in a brilliant, unforgettable flash of horror, September 11th, 2001 burned itself permanently into our psyches. As the twin towers collapsed, so did the economy. Travel budgets evaporated, training dropped precipitously in the hierarchy of corporate priorities, and as airplanes fell from the skies, airline fortunes followed as fear consumed the traveling public. The fact that corporations did not want to pay for travel became an academic concern because employees did not want to get on airplanes anyway.

Every cloud has a silver lining, even one as dense, dark, and impenetrable as September 11th. Literally overnight, videoconferencing providers saw demand for usage climb 300 percent. Web-based collaborative meeting services watched as requests for their applications spiraled upward. CBT, streaming media, and web-based university degree programs bathed in the adulation of thousands as recently reorganized priorities thrust them into prominence as viable alternatives to traditional services with a travel component.

It is now March 2002, and concerns about travel have abated as security at airports has increased. People are flying again, and just as important, the economy seems to have reached its low point and is turning around. Demands for training are starting to appear, and restrictions on travel have for the most part been lifted. Yet demand for videoconferencing, audio conferencing, web-based conferencing, and multimedia training remains high—very high, in fact. This time, it appears that these are here to stay. Usage may decrease somewhat, but the idea that they are "vanity technologies" that will fade away in favor of traditional offerings is wrong. Call it critical mass, serendipity, timing, or a fortunate alignment of the planets. The bandwidth is here, the fast processors are here, the costs are reasonable, the applications work, and most important of all, the sociological factors favor acceptance of alternative delivery media. If a person or company can accomplish the same thing via a videoconference or web session that they would otherwise accomplish by getting on an airplane, they will do so, with far less expense, human wear-and-tear, and greater efficacy.

About This Book

I originally envisioned this book as a study of videoconferencing: how to use it effectively, how to make it profitable, and when *not* to use it. However, as I began to examine the business of conferencing, I realized that there is much more to it than video. I realized that there is a whole family of services out there that can create enormous competitive advantages and dramatically reduce the cost of doing business for corporations if they are used properly. I also realized that many of the service components—the origination devices, the networks that transport the signal, and the devices that receive it—are mysteries to many people. Even the services themselves are often poorly known. For example, most people have heard of streaming media, but relatively few know what it really is, how it works or how it can be used, even when they are looking at it. This book, then, is designed to be a tool for companies and individuals who want to use video and audio conferencing techniques to their advantage. Although the book covers video, audio and network access and transport technologies in some detail, it isn't really a technology book. There are already many fine books on the market that cover these technologies in great detail, and there is simply no value in re-creating them here. So my intent is to give you enough technology to answer basic questions and enough knowledge to be able to seek more elsewhere if you decide you need it. What I *will* discuss in detail are the applications that these technologies lend themselves to, the considerations that must be taken into account before making decisions to use them, and the practical know-how that makes the difference between a successful event and a memorable and embarrassing failure.

This book is divided into three sections. Section One, "Conferencing Applications and Technologies," begins with an overview of the drivers behind the growth of conferencing technologies before describing typical business challenges and the conferencing applications that can resolve them. It presents and defines the many forms of conferencing before introducing the technology that makes it all work. Beginning with the origination equipment, we describe the typical network, the fascinating history of video, the layout of a video studio and videoconference room, and the equipment found there that makes the conference successful.

Next we discuss the anatomy of a video signal, looking at the alphabet soup of acronyms that characterizes the industry before introducing and discussing the access and transport technologies that empower a network, followed by a discussion of the signal termination equipment (television and PC).

In Section Two, "The Videoconferencing Industry," we examine the companies that make up the major segments of the conferencing industry and discuss the corporate considerations involved in using their services.

In Section Three, "Conducting a Successful Conference," we delve into the complex art of planning, preparing for, and conducting a live event. This section covers such critical topics as preparing the facilitator, creating video-ready materials, using studio equipment, and the roles of the various people who are involved in the process of conducting the media event.

ABOUT THE CD

If you browse the technical section of your favorite bookstore, you will find plenty of books that have a CD glued into the back cover. If you are like me, you have bought some of these books *because* of that CD, thinking that whatever was on it would bring great additional value to the reading experience, only to discover that it contained little of real substance—color versions of the black-and-white line drawings scattered throughout the book, for example. Zowee.

This book/CD combination is different. First of all, you have no doubt noticed that the CD is attached to the inside of the back cover. This is a book about videoconferencing, so we thought it only appropriate that we present the first section of the book to you in video format. So before you get too deep into the printed matter, open the CD and watch the video that you'll find there. It sets the stage for everything that follows in the text.

Second, because this is a book about (among other things) video, which is highly dependent on the richness of visual media, the gray scale photographs that appear in the book simply don't serve to illustrate the points that need to be made. So although you will find plenty of gray scale images and line drawings used throughout the book, they are also included in full color on the CD. I encourage you to refer to them for maximum visual impact and understanding.

Thank you for purchasing this book. I hope you find it useful and hope that you will take the time to drop me a line if you have a suggestion that will make the next edition better or if you simply want to chat. Enjoy—or perhaps I should say, "Action!"

STEVEN SHEPARD
FEBRUARY 2002
WILLISTON, VERMONT; MADRID, SPAIN; AND BUCHAREST, ROMANIA

PART

1

Conferencing Applications and Technologies

Why Conferencing?

Several factors contribute to the growing interest in video and other conferencing techniques. The first of these, sad to say, is that despite all claims to the contrary, many people are still justifiably nervous about getting on airplanes following the tragedies of September 11th. As a result, training classes and off-site meetings have declined dramatically.

The second (but equally important) factor is cost. Because of the overall decline in the economy, companies are less inclined to pay for travel expenses. Meetings and training events are clearly important —no company would ever say otherwise. However, if there is a different way to accomplish the same goal without having to pay for travel, lodging, and meals—in addition to the cost of an instructor if training is the goal—companies will sit up and take notice, particularly if the alternative is effective. Furthermore, because of the economy and the growing competitive nature of the marketplace, company management is less willing to have employees away from their jobs for long periods of time because of lost opportunities, especially in sales organizations. In addition to the real, measurable costs associated with travel, meals, and lodging, there is a less tangible (but no less real) cost associated with having sales professionals and others away from their jobs. Consider the following numbers: 20 salespeople (a standard class size) enrolled in a 3-day course results in 60 days (2 months) of lost sales effort. No matter how good or important the training or how crucial the meeting, managers must consider the cost of lost opportunity.

Another factor is that the alternatives now actually work. Alternative media, under certain carefully controlled situations, can do an effective job. It wasn't all that long ago when videoconferences were grueling tests of a person's patience. The quality of the image and the audio was so inferior that participants spent more time poking fun at the medium than they did engaging in the meeting.

Videoconference participants today are a jaundiced group because they have grown up with high-quality television that is colorful, sexy, richly contextual, and, for the most part, dependable. Like the telephone, it is always there and it always works. Videoconferencing technology, however, does not enjoy the same history. It has not been around as long as television, it does not generally have the dedicated bandwidth available to it that television does, it is not professionally directed and produced like television, and it is not designed for entertainment purposes. Participants invariably compare it to what they know and conclude that because it is not as good as television, it is not worth using as an information conveyance mechanism. This, of course, is flawed logic because the two media have very different charters. Although costs for conferencing facilities have come down, video and audio *Coder/Decoders* (CODECs) have improved, bandwidth has grown, signal termination devices (PCs and TVs) have become far more capable, and the software that decodes the video and audio signals has evolved dramatically, conferencing technologies still face an uphill battle. Interestingly enough, the issue is not technology based—it is a human-perception problem, and it must be overcome if conferencing is to become a mainstream service. Of course, the problem is not universal. Many large corporations have adopted video-conferencing and some use it globally. Richard Parlato, the founder of Proximity Inc. (www.proximity.com), one of the country's largest videoconferencing service providers, is quick to speak of the value of videoconferencing. "This is not a passing fad: videoconferencing is real and it's here to stay. Since the horrors of September 11th we have seen our business double in volume and it has not gone down—nor do we expect it to. We have large corporate customers—*multinational* corporations, mind you—that schedule conferences with us every day. It isn't a special thing: *this is how they hold meetings*.

"Our job is very simple: We remove distance as an issue. We provide a service that hides the complexity of videoconferencing technology from the customer so that they can concentrate on the task at hand—training, a company meeting, whatever. Managing the technology is our job, not theirs. If we perform our job properly, the customer is almost unaware that he or she is participating in a videoconference—the technology is that good today."

Parlato's observations are accurate. In the mid-1990s, an advertisement appeared in newspapers across the country. The ad was highly graphical and contained a single, large image: two hands clasped in a handshake. Below the hands was the following caption:

"How do you transmit a handshake?" The ad, placed jointly by a collection of major airlines, was a shot across the bow of the nascent videoconferencing industry. In 2001, Donna Mikov, the Vice President of Communications for Boeing Commercial Airplanes, made the following observation during an interview: "You can't fax a handshake, or e-mail the smell of a holiday dinner, or substitute videotape for the warmth of a grandparent's hug." But, of course, videoconferencing doesn't target those wonderful applications, nor should it. It is a business tool—nothing more and nothing less. Remember Richard Parlato's comment: "We remove distance as an issue." If conferencing technologies can accomplish that, they achieve their primary goal of making a difference in the cost of doing business. Enterprises apparently agree. According to a recent report from Frost & Sullivan, U.S. videoconferencing service revenues, measured at $1.48 billion in 2000, will grow to $5 billion by 2007. Wainhouse Research, meanwhile, projects growth in videoconferencing infrastructure spending to from $375 million to $1.25 billion in the same time period.

Conferencing Modes

There are three common conferencing modes: one-to-one, one-to-many, and many-to-many. One-to-one mode is typically used for a conversation between two individuals or between two small groups of people. One-to-many mode, the most common technique, is used for the widespread delivery of a common message such as in video-delivered training. It is sometimes called *point to multipoint*. Many-to-many mode is the least common form of conferencing. It requires video origination equipment at every site and a complex bridging arrangement to ensure that every site can transmit signals to and receive signals from every other site.

Related to these three modes are the actual service types that they provide. One-way audio/video is used to convey a common message when there is no need for audience feedback. One-way video with audio return only is used for situations when there is no need for the video originator (such as an instructor or meeting facilitator) to see audience members, but when he or she must be able to hear them.

Finally, two-way audio/video is used for fully interactive sessions that require complete two-way connectivity between all parties.

Conferencing Applications

Conferencing applications have evolved over time to address the challenges of a multitude of business scenarios. In this section we will present a series of these scenarios and the conferencing applications that resolve them. Remember that the intent of all conferencing applications is to eliminate the impact of distance by combining image and sound creation technology, image and sound transport technology, and image and sound delivery technology to bring about what many people call "the death of distance." If you think that videoconferencing technologies apply themselves only to meetings, consider the following examples.

Security

A multistate power consortium has power-generation facilities in six New England states. These facilities include nuclear- and hydroelectric-generation plants, two windmill farms on remote mountain ridgelines, a collection of cogeneration facilities, an experimental solar array, power delivery substations, grid-monitoring locations, and two network management centers for the optical fiber that resides within the hollow core of the aluminum power cables. Needless to say, security is a serious concern on many levels. At the most fundamental level, the company must ensure uninterrupted service for its hundreds of thousands of customers. Similarly, the company must be vigilant for signs of sabotage. Here is the challenge: how can the company's security organization closely watch not only the large, easily accessible power plants, but also the remote, high-tension towers and research facilities? They do not have enough personnel to physically inspect every facility on a continual basis, and even though they have a rapid response agreement with local police in their areas of operation, they must still detect and identify problems before calling the police for a dispatch.

The solution to this challenge is one-way video. In each state, the company established a highly secure network management and monitoring center with two responsibilities: to monitor the health of the optical fiber-based telecommunications network that is coaxial to the power grid and to provide surveillance of critical facilities. The major locations—generation plants, substations, and cogeneration facilities—are monitored by groups of carefully placed cameras that can be moved on gimbals to provide total facility coverage. Remote locations are monitored by carefully placed, solar-powered, slow-scan video cameras that capture single images every few seconds and transmit them to the network management centers. All cameras have infrared capabilities and can therefore capture images in complete darkness. Controllable cameras are mounted atop the tallest power towers to scan multiple towers in the chain.

From a single location, the entire network and all its facilities can be visually monitored, and between the company's security personnel and local police forces, any threat or service interruption can be dealt with quickly.

Medical Imaging

Northern Georgia has a large and widely scattered population base. There are many small towns, but few of them have enough critical mass to justify the establishment of a large hospital with significant specialization. As a result, patients who require more than just routine treatment must travel to a major city on their own or via medical transport, or must wait for diagnosis until a specialist visits their rural healthcare facility. These options are problematic for several reasons. First, many patients might be too ill to travel or might not want to travel for any number of reasons. Second, it may be logistically difficult for medical specialists to travel to remote facilities on a regular basis. Finally, small rural healthcare facilities might not have access to specialized diagnostic equipment typically found only in big city hospitals.

The solution came with the arrival of telediagnosis capabilities. Using a highly secure broadband facility, specialists at a major medical center can connect to medical-imaging devices located at rural hospitals and perform patient diagnoses as effectively as if they were face to face with the patient. Using videoconferencing techniques, they can conduct one-on-one interviews with both patients and primary caregivers, perform real-time monitoring of vital signs via telemetry devices, and even perform tactile physical examinations using highly specialized haptic devices. Telemedicine has become a multibillion-dollar business and is responsible for the availability of significantly enhanced medical care in rural areas throughout the world.

Telepresence

The term *telepresence* may be right out of Marin County, but the concept relates to pure big business. The prefix *tele-* is a Greek word that means "at a distance." Telepresence, then, means "presence at a distance," an oxymoronic expression but one that makes sense to anyone who has ever taken part in a videoconference. When it isn't possible for a participant to be physically present at a meeting or training class because of cost, travel anxiety, or a scheduling conflict, conferencing technology can create a virtual presence that eliminates the distance barrier. The use of this application has grown steadily in the last 10 years and has experienced exponential growth since the attacks of September 11th. Sprint saw a 40 percent growth in usage in the weeks following the attacks, and corporations around the world have begun to use it in earnest. Like many technology-based solutions, videoconferencing has routinely cycled between periods of high usage and relative obscurity. This time, however, it appears that the technology is here to stay for the long term because of relative acceptance by the right people, plummeting costs, widespread bandwidth availability, and highly capable, low-cost, *customer-provided equipment* (CPE) such as roll-around videoconferencing units, stand-alone cameras, ancillary devices, and CODECs.

Training

As part of the ongoing professional development process, a major multinational law firm with offices in 22 North American and European cities must conduct update training sessions for all partners about the current legal issues associated with sexual harassment. There are more than 350 personnel involved in all 22 cities, and a certified subject matter authority in the field must deliver the 4-hour training. The cost and logistics associated with conducting 22 separate sessions on 2 continents are deemed by the firm to be prohibitive. Professionally, however, they must comply with the mandate to maintain knowledge currency among all practitioners on a very sensitive legal issue.

The solution is distance learning. The subject matter expert, who lives in the northeastern United States, delivers the material from a videoconferencing studio in Burlington, Vermont. The videoconference services company arranges for a point-to-multipoint connection, which delivers the signal in real time to all 22 locations simultaneously. A sophisticated audio bridge enables participants to ask questions and carry on conversations between sites and enables the facilitator to conduct interactive case studies and what-if scenarios. The cost to conduct such a session is far less than it would be for a live session at each location and it takes far less time. Furthermore, every employee receives *exactly* the same message at *exactly* the same time, eliminating the possibility of misinterpretation of such critical subject matter. Finally, videoconferencing is much less invasive and disruptive than a typical classroom environment where participants must often travel to a remote training location. With videoconferencing, participants walk down the hall, take part in the conference, and return to their desks afterward with minimal disruption. Not too long ago only the largest companies had videoconferencing rooms on their premises. Today, the price to equip such a room has plummeted. A few years ago, a videoconference room was a $100,000 proposition. Today, a fully equipped room can be provisioned for less than $20,000, while a desktop device for simple IP conferencing can be purchased and installed for less than $100. Richard Parlato of Proximity observes that "companies, even *small* companies, have realized that they can install a videoconference room for the cost of three employee business trips. Once it's paid for, the only ongoing expense is the actual cost of the call, which in today's bandwidth market is ridiculously inexpensive. And if they register their room with us, they can even convert the room into a profit center." Proximity has a unique business model. In addition to their own rooms in Burlington, they have a business arrangement with hundreds of clients around the world. "Let's assume that your company, located, perhaps, in Miami, goes to the trouble and expense of installing a videoconferencing room. You register your room with us, and we enter your room and your business requirements into our database. If a client calls us and needs a room in Miami from 10 until 12 on Thursday, we call you and ask whether your room is available for a conference at that time. If it is, we arrange with you for entry of the participants into your facility at the appropriate time. We charge them a reasonable fee for the conference, manage the entire event for everyone concerned, and pay you a percentage of the fee, which covers your costs of maintenance and use. Everybody wins. It's a perfect arrangement for all concerned."

Collaborative Design

A naval engineering firm in San Diego that designs supertankers works closely with a shipyard at Hunter's Point in San Francisco, where their designs are converted from blueprints to vessels. Although the design firm has engineers and architects on site in San Francisco during the construction of a vessel, they also have staff members in San Diego who must monitor the various phases of construction for liability and quality-control purposes, including 3-D image overlays as the hull's construction progresses to ensure design compliance. The firm uses a variety of imaging techniques to do this, including the transmission of still images of the hull from various angles, captured by cameras mounted around the construction site. They can also transmit slow-scan video from the construction site. The use of these technologies dramatically reduces the engineering firm's travel expenses and enables the home office to perform real-time monitoring of ongoing projects.

Scenario Planning

The situation was tenuous and getting worse. A large construction firm with offices throughout the world finds itself facing the prospect of having to declare *force majeure*[1] in Indonesia because of the threat of encroaching rebel activity in the area. Because of its long-time multinational presence, the firm has successfully faced similar challenges in the past.

Before making what would amount to a largely unilateral decision, the in-country management team asks for a meeting with senior managers around the world as well as past country managers who faced and dealt with similar challenges during their own tours. Using a hybrid service available through a variety of companies, a conference is scheduled. Participants call into an audio conference bridge while simultaneously connecting to a secure web site that is only available to them. Via the audio bridge, participants can talk freely with each other, while the web site gives them the ability to share graphical information such as maps, meeting agenda items, and other data critical to the meeting. Because of the ability to meet in real-time and share current and past knowledge, the company may be able to craft a solution that does not involve shutting down operations, as *force majeure* situations often demand.

Video Postproduction

A video production company has completed the rough-cut assembly of a nature film for Animal Planet and now needs to have it reviewed by the sound engineers, the content experts, and the continuity people. Unfortunately, the film assembly was done in the United Kingdom, while the people who need to review it are in three different cities scattered across the United States. One solution is to send time-code burn-in copies to everyone and have them review the film

---

[1]Literally, this means "*major force.*" This is typically described as an unexpected or uncontrollable force that results in a cessation or a severe curtailment of business operations.

and forward their comments to the director. However, this would take far too long and would obviate the advantage of having everyone share their comments with each other in real time.

The solution, of course, is videoconferencing. Using a relatively high bandwidth connection, the production house arranges with a videoconferencing provider to establish a point-to-multipoint connection between the production facility and the location of each of the participants. The video is transmitted to each site simultaneously, and each location has the ability to share comments over an audio bridge with everyone else. In some cases, it may be possible for each site to have video control so that they can stop the program, back up, or fast-forward as they desire in order to make a point. This solution saves time, travel money, and the cost of misinterpretation.

Just-in-Time Knowledge

A molecular biology laboratory in Melbourne, Australia, has recently purchased a complex and extremely expensive device from a Boston-based company used to perform biochemical assays of complex molecules. Now, they need to install an add-on device that will improve the accuracy of their work, but for some odd reason the device will not attach to the primary device as the instructions indicate. They have a choice: they can wait several days for an expert from the manufacturer to arrive from Boston, a delay that is unacceptable given the project timeline they are working under, or they can conduct a videoconference with the manufacturer's subject matter expert so that he or she can walk them through the problem in a matter of minutes.

Architecture

An architectural firm designs a complex and high-tech office building for an advertising client. Light is critical to this particular client because of the artists they employ, so the client wants to make sure that the building's design permits the proper levels of light to enter the spaces during all four seasons of the year. It would be prohibitively expensive—and probably impractical—to make changes once the building is complete. Yet long before the first shovel breaks ground at the building site, the client has walked through the building, rearranged furniture, adjusted the size and angle of windows, and widened a staircase. How? With an immersive virtual-reality application that creates a 3-D model of the building that the client can walk through using a special viewing device. When the building is complete, it is exactly what the client wants—there are no surprises.

Military Surveillance

Military threats have been made by rebel forces against local civilians in a small central African country. The country's government is strongly supported by the United States, but the local government feels that it would be politically imprudent for the United States to send in relief forces at this time. Instead, the U.S. military agrees to provide surveillance capabilities. Using high-resolution imaging satellites, the National Reconnaissance Office provides real-time images of the threatened area and warns government forces of rebel troop movements that could indicate an impending strike against civilian targets. This enables the local army to muster troops and move them to the appropriate place in anticipation of an attack.

Telesurgery

A patient at a remotely located research facility suffers a severe gall bladder attack, and the local physician determines that the condition requires surgery that he is not qualified to perform. The patient's condition is deteriorating rapidly, and there is not enough time to transport him to a treatment facility. The solution is telesurgery, a new technique that is actually being used today under carefully controlled conditions. (In 2001, a patient in France had his gall bladder removed remotely by a surgeon physically located in the United States.)

With telesurgery, the patient lies inside a special operating theater device that is connected to the remote surgeon with a high-speed network facility. Inside the remote surgery environment are robotic devices that can be controlled by the remote surgeon and that can be modified to carry a variety of surgical tools. The surgeon inserts his or her hands into special gloves and looks through a high-definition viewing device that provides a 3-D view of the actual operating area. When the surgeon moves his or her hands, the remote device moves at the same time and exactly the same distance, but this time the device is actually performing a surgical procedure. Latency is less than 200 ms.

Another capability that virtual-reality surgery uses is a technique called *haptics*. Haptics is the science of providing remote tactile feedback for applications such as telesurgery, aviation, and other applications that require a sense of touch. For example, when the tiny remote suture needle pops through the skin, the surgeon feels the needle pop through the skin, several thousand miles away. This technique, a form of virtual reality, is actually becoming a reality today.

One-to-Many Audio Distribution

A corporate executive wants to deliver an important speech to every one of his 55,000 employees worldwide. Obviously, there is no way to bring the entire corporation together for a real-time presentation. The alternative is to tape the speech the first time, digitize it as MPEG-2, and place it on a streaming media server, thereby making it available to all employees at their convenience. For the next few weeks, the executive makes himself available on a conference bridge twice a week to answer questions.

Digital Movie Distribution

The movie *Bounce*, starring Gwyneth Paltrow and Ben Affleck, was the first feature film to be distributed digitally to certain theaters—that is, no reels of film to be shipped. This technique will undoubtedly become the standard for film distribution in the future for two reasons: it reduces distribution costs for the producer of the film and places more control into the hands of the theater manager. Studios will no longer have to pay the cost of manufacturing thousands of copies of a film and shipping the copies around the world, and theater managers will enjoy greater control over their own business. For example, suppose a 10-theater multiplex is showing 8 different movies. One of the movies has been open for several weeks and attendance has dropped off. A new release, however, is so popular that the theater is turning people away. With digital film, the theater manager can show the film in an additional auditorium on the fly, making room for people who want to watch the new film.

These are only a few examples of the business challenges that can be resolved through the appropriate use of audio or videoconferencing technologies. We now turn our attention to the various forms of conferencing that are in use today. In this section, we will describe the characteristics of each one, including the applications at which each is targeted and the degree of complementarity that each has with peer applications.

Conference Techniques

Seven variations on conferencing will be discussed in this section. Many are technically similar so it is important to understand the functional differences between them. The services covered include videoconferencing, teleconferencing, IP conferencing/multicasting, streaming media, web conferencing, imaging, and desktop conferencing.

Videoconferencing

Videoconferencing takes on a variety of forms depending on the scenario at which it is targeted. Most commonly, it is used to facilitate point-to-point or point-to-multipoint meetings among individuals located far from each other. For example, a manager who has employees in three cities may want to conduct a staff meeting, but does not want to incur the costs of travel for so many people. As Figure 1-1 illustrates, each location has a videoconferencing room that

Management Console

Figure 1-1

Figure 1-1

Multiple sites
connected via a
video bridge

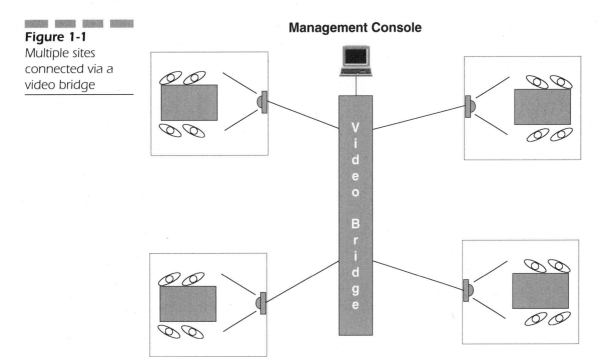

houses a videoconferencing unit. The unit comprises one or two monitors (one to display the local room and the other to display whichever remote room "has the floor") and a video camera that can be adjusted to follow the movements of whichever person is talking. In some advanced units, the camera automatically follows the voice of the speaker. This type of setup enables each location to see other locations. In most cases, the system is set up in such a way that whichever site is generating the voice signal (talking) is displayed on all remote monitors. As soon as another person at a different site interrupts, the conferencing equipment automatically switches the view to that room.

Unless the conference is point to point (two sites only), a video bridge is involved in the establishment of the conference. The video bridge enables multiple locations to be simultaneously connected to one another and permits the videoconferencing service provider to monitor the call for *quality of service* (QoS) assurance purposes from a single management point. A good example is the *Indiana Higher*

Education Telecommunication System (IHETS). IHETS has two *Asynchronous Transfer Mode* (ATM) multimedia conference units (MCU video bridges): one with 32 ports and the other with 40 ports. Point-to-point calls are not scheduled using the bridge. Typically, conference bandwidth is 384 Kbps; remote units must be *Integrated Services Digital Network* (ISDN)-capable to function properly with the ATM-based bridges.

A typical corporate-level videoconferencing unit such as the one shown in Figure 1-2 incorporates an *inverse multiplexer* (IMUX) that combines the bandwidth available in three or more ISDN *Basic Rate Interface* (BRI) lines to create a super-rate, two-way channel for transporting video and audio signals, as illustrated in Figure 1-3. Three ISDN BRI lines provide six 64 Kbps channels for an aggregate channel of 384 Kbps, which is perfectly adequate for high-quality videoconferencing. These devices range in price from as low as $6,000 to as high as $80,000. Although many people would turn up their noses at the thought of using ISDN, let it be known that far and away the vast majority of all videoconferencing rooms use ISDN for network connectivity as opposed to some other protocol.

Figure 1-2
A Tandberg videoconferencing unit

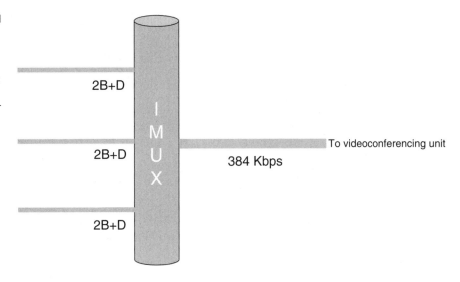

Figure 1-3
Three ISDN BRI circuits are bonded to form a 384 Kbps circuit.

Teleconferencing

A teleconference does not include the video component. Instead, participants dial into an audio bridge, as shown in Figure 1-4. The audio bridge ensures that levels are correct for all parties because audio volume tends to drop off as more people are added to the call. Although three-way calling is common today, most major service providers offer an audio-conferencing service that provides security, call management, and logistics administration.

IP Conferencing

IP conferencing is really no different from videoconferencing and audio conferencing, except for one thing: the conference is conducted over a network that uses the *Internet Protocol* (IP), such as the Internet itself. This is a growing area of interest today because IP has matured to the point that Internet-based networking can offer relatively high levels of service quality to users. Today, combinations of network technologies enable IP to provide very good video services at a low cost (50 percent less than ISDN, for example) to customers.

Figure 1-4
Multiple
participants
connected through
an audio bridge

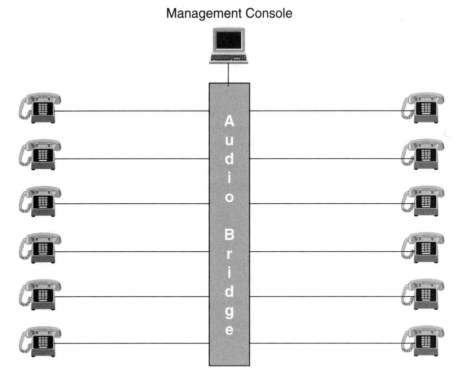

Management Console

There are several options for delivering IP conferencing. The level of quality distinguishes one from another. A conference conducted over the public Internet will be of questionable quality at best. Because there are no guarantees with regard to the route that the IP packets will take from the source to the destination, there are also no guarantees with regard to the end-to-end delay that will occur or with the jitter (variation in delay) that the packets will experience as they make their way through the network. For video service of any quality, public Internet IP conferencing should be avoided.

The next option is to use a corporate *virtual private network* (VPN). A VPN is a shared network with the capability to carve out bandwidth for a particular application, giving the users of that bandwidth the sense that they are on a dedicated facility. This technique can work well, but it can also be highly problematic. For example, if the IP VPN shares bandwidth with the corporate intranet/*local area*

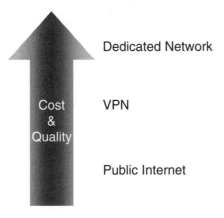

Figure 1-5
Cost versus quality
in network options

network (LAN), serious and unanticipated contention can occur between video traffic and other corporate traffic, dramatically reducing the quality of the final product. Most corporations that use IP videoconferencing extensively within the enterprise do not mix the two. They keep the LAN IP traffic separate from the video IP traffic.

The third alternative is to use a dedicated facility for IP packet transport. As Figure 1-5 shows, this is the most costly solution, but it also offers the highest level of repeatable service quality.

The IP protocol will be discussed later in the section "Internet Protocol (IP)" when we discuss the technologies themselves.

Streaming Media

Streaming media is a relatively new technology that is beginning to have a major impact on the enterprise and on network providers. Because it is relatively new and not yet well understood, we will spend some time on it here. Although it is new, it is already generating impressive revenue numbers. Streaming media was a $400 million market in 2001, including products and services. According to Gerry Kaufhold of Cahners InStat, service providers in 2001 enjoyed revenues from streaming media in the neighborhood of $900 million. An additional $734 million was spent on servers and technologies for streaming media support. Even though these are relatively small numbers by network service provider standards, Cahners InStat predicts that this same market will grow to $7.7 billion in 2005.

Lawrence Orans, an analyst with Gartner Group, estimates that 80 percent of all Global 2000 companies will routinely use live and on-demand video to the desktop by 2006.

Streaming media has a number of key benefits. One of them is immediacy. The user begins to see the streamed content within seconds of booting up the appropriate media player. This is important: those readers who remember the early days of multimedia delivery via the Web will recall that the process took significantly longer because the player had to wait until the entire file downloaded from the source.

There are numerous applications for streaming media. Corporations with multiple locations can use streaming media to deliver a common message to every employee in the global corporation. Those who are not able to view the broadcast when it is first conducted can stream it to their desktop and view it when they have time. A growing number of companies use streaming media to better explain their products to potential customers.

There are several well-known formats for the delivery of streaming media. These include RealMedia from RealNetworks, Windows Media from Microsoft, QuickTime from Apple, and the various flavors of MPEG from the Moving Picture Experts Group. RealNetworks and Microsoft have the greatest penetration, but Apple QuickTime and streaming engines based on MPEG also have significant followings. The big three—RealNetworks, Microsoft, and QuickTime—all rely on proprietary technology for encoding the media streams that they transport, requiring users to deploy proprietary servers, encoding software, and player technology. The good news is that all three companies provide free versions of their player software. This is one area where MPEG has an advantage: it is standards based and therefore universally accepted around the world.

How It Works There are four steps in the process of creating streaming media: capturing the content, editing it to the appropriate length, encoding it for transport, and delivering it across the network. The encoding and delivery processes have the greatest impact on network providers because they determine the bandwidth requirements.

Most users rely on the *Hypertext Transfer Protocol* (HTTP) for media delivery. However, if the streaming application is for a single corporation, for example, the IT staff may select a single format for their streaming media application and install format-specific servers that enable them to take advantage of certain features unique to the chosen format. For example, corporations will often choose RealServer or Windows Media because they are well deployed, well understood, fully supported, and readily available.

Microsoft relies on the *Microsoft Media Server* (MMS) protocol for communication between its delivery platform and proprietary player. Similarly, Apple and RealNetworks use a combination of the *Real-Time Streaming Protocol* (RTSP) and the *Real-Time Protocol* (RTP) for communication between the two. The only drawback to the use of these protocols is that some firewalls will block this traffic and prevent it from reaching its LAN destination if filter settings are established to do so.

RealServer can be run on UNIX, Linux, and Windows-based servers, and provides a 25-simultaneous-stream version of its server at no cost. Microsoft offers its unlimited user-streaming server as part of the Windows 2000 operating system.

Network Implications Streaming media can be either live or archived and then delivered on demand. Regardless of whether the data is to be delivered live or from an archival source, bandwidth is the key factor with regard to the quality of the delivered stream and ultimately the experience that the user will have. Audio and video packets, created from the original data stream, must be delivered sequentially and with minimal latency (jitter) if the quality of the delivered product is to be acceptable.

One way that streaming media applications reduce the impact of network congestion is through a process called *buffering*, which simply means that the media player collects and stores a certain volume of incoming packets before it begins to execute the file.

Consider the example shown in Figure 1-6. The diagram shows four buckets, each with a tube at the bottom that feeds the stream of incoming content to the user. In the first bucket, A, the user sends a request upstream to the streaming media server, asking that content be delivered. The media server responds in B, sending packets

Figure 1-6
Managing bursty
traffic in a
streaming media
environment

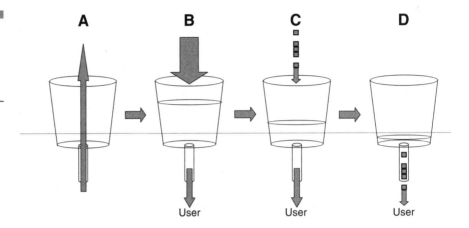

downstream to the user's PC where they will be assembled into the video or audio stream that the user anticipates. Before the media player begins to execute the file, however, it first goes through the exercise of collecting a certain volume of packets in the buffer bucket. Once it has accumulated the appropriate number of packets, it begins to execute the file, all the while continuing to accumulate packets in the buffer. At all times it must attempt to maintain the volume of packets in the buffer above the *low watermark* illustrated by the dotted line. This ensures that at a certain delivery rate there will always be enough packets in the buffer to maintain signal-delivery quality.

However, there is always the possibility that because of network congestion or some other anomaly, the flow of packets into the buffer from the network is choked off, as illustrated in C. As long as the flow reduction does not go on for too long, the media stream will continue to play correctly. If the buffer runs dry, however, the stream is interrupted and the user's video or audio will falter, sometimes stopping altogether until flow recommences and the buffer fills once again to the minimum acceptable level. This is shown schematically in D.

This technique is not exclusive to archived content. Live video is also buffered, a fact that typically results in a play delay of approximately 30 seconds between the time an event commences and the time the streamed content reaches the viewer. Some content providers who use QuickTime deliver content using a technique called *progressive download*, in which the content is partially downloaded before it begins to play.

Both live and archived content can be delivered using *unicast* and *IP multicast* technology. With unicast, a stream of content is delivered to each client on demand. The result is that if 5 users request 100 Kbps streams of content that will be delivered simultaneously, the server must be able to accommodate 500 Kbps of outbound bandwidth. This becomes problematic when extremely popular sites (Morpheus comes to mind) are besieged by hundreds of users, all trying to download multimedia content from a single server. The result is a massively overburdened server and network. This can lead to the equivalent of denial-of-service attacks because the unfettered inbound traffic causes the server to thrash.

In an IP multicast environment, on the other hand, streamed content is broadcast across a network from the source server as a single stream of IP packets. This is accomplished through IP-multicast-enabled routers, which transmit a single stream to each connected router. This process is repeated until the transmission reaches all clients who want to receive the broadcast content.

One downside of IP multicast is that there is no single protocol that defines how the protocol works. There have been several proposals developed since its inception in 1988, the most popular of which is *Protocol Independent Multicast* (PIM). Like so many well-deployed protocols today, PIM was originally created by Cisco and is now being adopted by the *Internet Engineering Task Force* (IETF).

PIM has two basic forms: dense mode and sparse mode. *Dense mode* floods the network with a broadcast so that every IP multicast router in the network receives the signal. Routers that do not have clients who have requested the signal are eliminated from the broadcast on the fly.

Sparse mode, on the other hand, only transmits the signal to those routers that request the content. A request from a client travels to the source media server, at which time a multicast connection is established between the client and the server. A variety of proprietary protocols such as the *Session Announcement Protocol* (SAP) enable servers to advertise programs that are available.

In summary, dense mode IP multicasting is good for networks with plenty of available bandwidth and a population of users who are evenly distributed across the network. Sparse mode, on the other hand, is better suited for networks with users who are more broadly scattered across the network.

For IP multicasting to work properly, every router in the network must have the capability enabled. Furthermore, IP multicast may present reliability issues. When it transmits content into the network, there is no feedback mechanism that enables a network manager or user to monitor QoS issues. Some servers that support IP multicasting have the capability to receive a reverse-channel connection from the player (and therefore content), a capability that could help with reliability management. However, this can also diminish network scalability because as the number of back-channel connections increases, the amount of available bandwidth decreases.

A number of protocols are being considered that will make the implementation of IP multicast more routine. These include the *Border Gateway Multicast Protocol* (BGMP) and *Multicast Address Set-Claim* (MASC). The goal is to make IP multicast work properly across large public networks such as the Internet; for now, however, it is best suited for private corporate implementations.

One way to reduce the impact of delay across large networks is to distribute the deliverable content across the network, making it available from various sources and reducing the total number of hops a stream of packets has to make to reach the destination server. Companies that use streaming media routinely build their own content delivery networks that include the caching capability.

Both RealNetworks and Microsoft rely on server technology that enables the delivery speed to be increased or decreased to compensate for variable network congestion. Alternatively, some content providers create files of different sizes (and quality) for each level of bandwidth and enable the end user to select the stream that is most appropriate for his or her access method.

Streaming media consumes significantly more disk space and network bandwidth than traditional *Hypertext Markup Language* (HTML) web pages, which translates into a high-end server with plenty of internal resources and a high-speed network connection—T1 or faster.

Content Creation   Before it can be streamed on the network, the content must be created. The first stage of the creation process is called *content capturing*. This is simply the process of converting audio or video content into digital format or (capturing the first generation of the content digitally). Using dedicated capture cards, a PC can convert the analog signal from an analog playback device into digital format, which can then be stored on a hard drive. This process can be as simple as plugging a microphone into the jack on the side of a PC and recording the speaker's voice using special capture software. For video, a digital video camera can output directly to the PC via an *Institute of Electrical and Electronics Engineers* (IEEE) 1394 interface or Firewire card.

Once the content is captured, editing tools can be used to make final adjustments to it. There are many such packages on the market today. It is good to know, however, that all of the major players offer free versions of their encoders to enable a beginner to experiment before making a large technology investment.

Network Functionality   There are two primary techniques used to deliver streaming content across the Web. The first relies on a standard web server that delivers the requested files to a PC-resident media player. The second technique uses a stand-alone streaming media server that is optimized for the delivery of streamed content. Although streaming from a web server can be an effective solution, a streaming media server is significantly more efficient and produces better end results. The process of downloading media files across the Web has historically been a straightforward process of requesting the file, waiting for it to download, and playing it on a media player. There was no streaming per se; it was a simple, traditional download. However, as content streaming has become more common, both techniques have become embedded. The first is to use a web server, as described earlier; the second is to deploy a streaming media server.

With the web server approach, uncompressed content (audio and/or video) is compressed into a single media file for transmission across the network. During the compression process, it is optimized for a certain level of network bandwidth. This file is then placed on a standard web server and linked to a web page that describes the media file. This page, when accessed, launches the client machine's player and downloads the file. Web-server streaming relies on the HTTP, the protocol used by all web servers for server-to-client communication. HTTP resides above the *Transmission Control Protocol* (TCP), which governs all data transfers and attempts to ensure QoS.

When a streaming media server is used, the initial stages of the transfer are similar to those used with a web server, except that the media file is archived on a dedicated streaming media server instead of a traditional web server. A web page with a pointer to the media file is then placed on a traditional web server.

Content Delivery   The media delivery procedure in a media-streaming environment is significantly different from that used with a traditional web server. Instead of simply *shotgunning* the content to the user, data from a streaming media server is transmitted to the client in a far more intelligent fashion, ensuring that the content is delivered at the proper data rate required by the compressed content stream. Because of the logical relationship that is created between the client and the server, the server can respond to client feedback as required. Even though streaming media servers have the capability to use HTTP and TCP, they can also employ more specialized protocols such as the *User Datagram Protocol* (UDP) to improve efficiency. UDP is a lightweight protocol that offers some of the capabilities offered by TCP but with much less overhead. As a result, UDP is ideal for the transmission of real-time audio and video.

The Realities of Streaming Media   Corporations looking to implement streaming media face numerous challenges. Some of the challenges are technical, whereas others are managerial in nature. The technical challenges are relatively easy to overcome, whereas the managerial challenges tend to be longer term and require greater effort:

■ *Determine how and where the streaming media will be hosted.*
Many enterprises rely on the hosting services of an *Internet service provider* (ISP), and although this may be the best solution for most companies, it will not work for all of them. For example, a large corporation with a dedicated and skilled IT staff that already hosts web content may be perfectly capable of managing the streaming server and its content.

■ *Solicit management buy-in for the plan to host video on the Web.*
Streaming media offers a number of advantages for the

enterprise. Video tends to increase web traffic, a major factor for corporations that depend on web traffic for contact with the customer. Furthermore, multimedia content (audio and video) is an extremely rich media for the delivery of information and tends to hold the attention of the customer much better than written text. If the content is training related or has to do with archival content from a speech, meeting, or executive presentation, employees can view the video or listen to the audio as their schedules permit.

- *Determine how the hosted content will be produced.* Although some large corporations have their own internal production facilities, most do not. As a result, many companies will have to hire freelance production teams to shoot and produce video and audio content.

- *Determine the encoding method that will be used for streaming media.* Most sophisticated production houses have two encoding workstations: one to produce live content and the other to produce archival content. Smaller organizations outsource the encoding function because it is more cost-effective to pay a specialist to perform the work than to hire a full-time employee for the same task, particularly if the production volume is relatively low. However, a low-end solution can easily be created by installing a video capture card in a PC. The entire system should cost less than $2,000.

- *Research the alternatives carefully, but pick one solution.* Whether you choose Windows Media Player, RealNetworks' RealPlayer, or some other solution is immaterial. What matters is that you select one platform and stay with it as long as it satisfies the needs of the corporation.

- *Plan to perform routine content updates.* One of the greatest traps that companies fall into is the trap of aging content. In today's marketplace, content becomes obsolete relatively quickly, and because of the visibility that a web site or streaming media server has, the age of the hosted content will reflect poorly on the company. Make it a priority to produce content that doesn't age quickly and constantly check the content for currency. If possible, designate one or more employees whose responsibilities include regular examinations of the hosted content for currency and accuracy, and give them the authority to initiate updates as appropriate.

Web Conferencing

For the most part, web conferencing is a bit of a misnomer in that its name implies that some form of video will be used across the Web, and this is not necessarily the case. The most common (and successful) forms of web conferencing are actually hybrids of various conferencing techniques, such as those provided by companies like PlaceWare, Centra, and Webex. With these services, clients schedule a conference with the service provider, who then helps them establish the conference. Participants are sent a toll-free number via e-mail with an access code, which will enable them to call into an audio bridge at the time of the event. They are also sent the URL of a web site where the content that is to be presented is archived, along with several layers of security to ensure that the material is only available to valid participants. At the time that the conference is to take place, participants dial into the audio bridge, enter their password, and log in. At the same time, they connect to the web site and log in as well. As soon as the conference begins, participants hear the session facilitator over the audio bridge and see the materials (typically PowerPoint) on their computer screen. The facilitator can highlight objects on the screen, draw circles and arrows and other forms, and control the slides. Participants can indicate to the facilitator that they have a question as well as ask the facilitator to speed up or slow down. This solution is quite popular and is an effective alternative as long as the session does not last longer than a few hours.

Imaging

Imaging is the technique used to deliver a still image from a source to a destination. Commonly used by architects, engineers, and medical personnel, imaging has one key advantage: because it is a still image, the technique does not suffer from jitter at all and latency is a minor issue. If the image takes a few more seconds to arrive at the final destination, so be it. Imaging is often combined with an audio-conferencing bridge so that participants can discuss the content of the image that is simultaneously sent to their PCs.

Desktop Conferencing

Desktop conferencing is really nothing more than a variation on a theme. Instead of terminating the signal on a television or videocon-ferencing unit, the signal is presented to a PC, where it is displayed on the monitor. Many corporations talk a lot about this technique, but in reality, few have actually implemented it because of QoS issues and because it tends to share bandwidth with the corporate LAN and can have deleterious effects on the delivery of other network traffic. Telecommuters and *Small Office/Home Office* (SOHO) workers use desktop conferencing, but again, the volume of calls is relatively low.

Video Technology

It is now time to shift gears and discuss video technologies such as source devices, transport networks, and termination devices. We begin our discussion with a high-level examination of the overall net-work; please see Figure 1-7.

We begin with the source of the video content. A video camera (A) or videoconferencing unit captures the incoming video and audio sig-nals and carries those signals across an access network (B) to a net-work switch, bridge, or server (C). The access network may be dial up, ISDN, cable, *Digital Subscriber Line* (DSL), wireless, or a dedicated connection. The server transmits the data, now either an uninter-rupted stream of bits or a series of packets, into the transport net-work (D). If B is a switch, it will establish an end-to-end path between the two communicating endpoints. If it is a video bridge, it will pro-vide the audio and video connectivity among two or more locations. If it is a server (E), it will most likely be used to stream archival content across the network to users on demand, although not necessarily—

Figure 1-7
A typical video
network

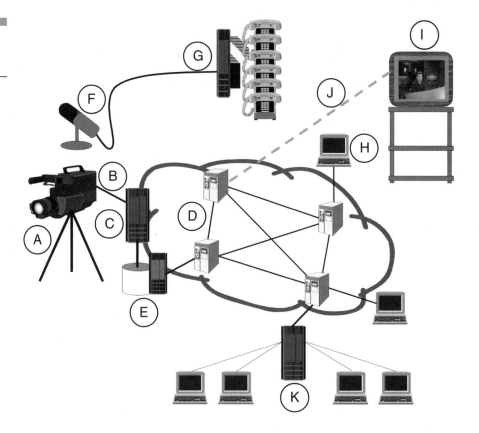

live traffic can be streamed, but the delays that are inherent in streaming traffic across a public IP network are well known.

At the same time that the videoconference is taking place, the speaker's audio may be captured (F) and transmitted to a secure audio bridge (G) so that users who do not have access to a videoconference unit can still take part in the broadcast. Those users, for example, may view the PowerPoint slides on a PC (H) while listening to the facilitator's voice over the audio bridge.

The transport network's job is to receive the traffic from the access network and move it to the destination as quickly as possible while preserving the integrity of the transmitted signal. Many different options exist here: the *Public Switched Telephone Network* (PSTN), frame relay, ATM, point-to-point T1/E1, and even optical solutions such as *Synchronous Optical Network* (SONET) or *Synchronous Dig-*

ital Hierarchy (SDH). All of these will be discussed in some detail later in this section, but for now suffice it to say that each offers certain advantages.

On the receive side of the network, there are several options as well. If the content is broadcast across a videoconferencing system, it will be received by a stand-alone or PC-based videoconferencing unit (I) via another access network (J). As before, the access network may use any of a number of technologies including dial up, ISDN, cable, DSL, wireless, or dedicated private line. These may also be terminated at a PC if the content is streamed via a web server or streaming media server (K).

These diverse technologies are combined to create a network with the capability to transport a relatively high bandwidth video signal with embedded audio from a source device to one or more destination devices while preserving the integrity of the video signal. This signal is remarkably complex, and in this next section we will describe its fascinating history as well as the technologies on which it is based.

Video: A Perspective

During the last few years, a revolution has taken place in visual applications. Beginning with simple, still-image-based applications such as grayscale facsimile, the technology has diverged into a collection of visual applications that include video and virtual reality. Driven by aggressive demands from sophisticated, applications-hungry users and fueled by network and computer technologies capable of delivering these bandwidth-and-processor-intensive services, the telecommunications industry has undergone a remarkable metamorphosis as industry players battle for the pole position.

Why the rapid growth? Curt Carlson, Vice President of Information Systems at the Sarnoff Corporation, observes that more than half of the human brain is devoted to vision—an indication that vision is our single most important sense. He believes that this rapid evolution in image-based systems is occurring because those are the systems that people actually need.

"First, we invented radio," he observes, "then we invented television. Now we are entering what we call the age of interactivity, in which we will take and merge all of those technologies and add the element of user interaction. Vision is one of the key elements that allow us to create these exciting new applications." Indeed, many new applications depend on the interactive component of image-based technologies. Medical imaging, interactive customer service applications, and multimedia education are examples.

Still Images

Still image applications have been in widespread use for some time. Initially, there were photocopy machines. They did not provide storage of documents nor did they have the capability to electronically transport them from one place to another. That capability arrived on a limited basis with the fax machine.

Although a fax transmission enables a document to be moved from one location to another, what actually moves is not document content, but rather an *image* of the document's content. This is important because there is no element of flexibility inherent in this system that lets the receiver make immediate and easy changes to the document. This capability requires that the image first be converted into machine-readable form, a capability that is just now becoming possible with *Optical Character Recognition* (OCR) software.

As imaging technology advanced and networks grew more capable, other technological variations emerged. The marriage of the copy machine and the modem yielded the scanner, which enables high-quality images to be incorporated into documents or stored on a machine-accessible medium such as a hard drive.

Other advances followed. The emergence of *high-quality television* (HQTV) coupled with high-bandwidth, high-quality networks led to the development and professional acceptance of medical-imaging applications, with which diagnosis-quality X-ray images can be used for remote teleradiology applications. This made the delivery of highly specialized diagnostic capabilities to rural areas possible, a significant advancement and extension of medicine.

Equally important are imaging applications that have evolved for the banking, insurance, design, and publishing industries. Images convey enormous amounts of information. By digitizing them, storing them online, and making them available simultaneously to large numbers of users, the applications for which the original image was intended are enhanced. Distance ceases to be an issue, transcription errors are eliminated, and the availability of expertise is no longer a problem.

There are downsides, of course. Image-based applications require expensive end-user equipment. Image files tend to be large so storage requirements are significant. Furthermore, because of the bandwidth-intensive nature of transmitted image files, network infrastructures must be reexamined.

The History of Video

Video has a long and colorful history that begins in 1951 at RCA's David Sarnoff Research Institute (now the Sarnoff Corporation). During a celebration dinner, Brigadier General David Sarnoff (see Figure 1-8), Chairman of RCA and the founder of NBC, requested that the institute work on three new inventions—one of them was called a *videograph*. In his mind, a videograph was a device capable of capturing television signals on some form of inexpensive medium, such as audiotape.

Figure 1-8
Brigadier General
David Sarnoff

Don't forget the time frame: thanks to Philo T. Farnsworth (yes, that's his real name), who, among other things, invented the predecessor of today's *cathode ray tube* (CRT), electronic television became a reality in the 1920s. Farnsworth was an amazing man; a slight digression about him is worth taking. Farnsworth, whose picture is shown in Figure 1-9, was born in Beaver, Utah, and was educated in the Utah and Idaho public school systems. While attending high school in 1921, he studied the molecular theory of matter, the behavior of electrons, and the special relativity theories proposed by Albert Einstein. He also studied the inner workings of automobile engines and various forms of chemistry. During this period, with no formal knowledge of electronics, he sketched his concept of electronic video on a chalkboard (see Figure 1-10) for his science teacher. He was 14 years old. Years later, he would fight (and win) a protracted patent battle with RCA over the invention of the technology.

Farnsworth eventually enrolled in Brigham Young University, but left the university at the end of his second year when his father died. In 1926 he joined the Crocker Research Laboratories in San Francisco, and soon produced the first electronic television image. Crocker Research Laboratories later became Television Laboratories, Inc. and in May 1929 became Farnsworth Television, Inc.

Figure 1-9
Philo T. Farnsworth

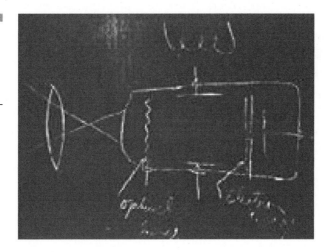

Figure 1-10
Farnsworth's first illustration of the concept of television

Farnsworth's television research covered a broad array of issues associated with the technology, including image scanning, focusing, signal synchronization, image contrast, image controls, and unit power. He also created the first CRTs and the first simple electronic microscope. Farnsworth experimented with the use of radio waves to determine direction (later named *radar*) as well as the use of black light for night vision.

During the 1960s, he worked on special-purpose television projects and numerous initiatives associated with the peaceful uses of atomic energy. At the time of his death at age 64, Farnsworth held more than 300 patents and was 1 of 4 inventors honored in 1983 by the U.S. Postal Service with the issuance of a stamp bearing his portrait.

By the early 1950s, black-and-white television was widespread in America. The gap between the arrival of television and the demand for video, therefore, was fairly narrow.

Work on Sarnoff's videograph began almost immediately and a powerful cast of characters was assembled. There was one unlikely member of this cast who served as a catalyst: Bing Crosby. Keenly interested in broadcast technologies, Crosby wanted to be able to record his weekly shows for later transmission. The Bing Crosby Laboratories played a key role in the development and testing of video technology.

The Sarnoff Institute called upon the capabilities of several companies to reach its goal of creating what Sarnoff dubbed the *Hear-See Machine*. One of them was Ampex, the developer of the first commercial audiotape recorder. Because of its magnetic nature, the Sarnoff team believed that audiotape technology could be applied to videorecording.

To a certain extent, they were correct. Marvin Camras (see Figure 1-11), a Sarnoff team member and scientist who developed the ability to record audio signals on steel wire used during World War II, soon discovered that the video signal was dramatically broader than the relatively narrow spectrum of the audio signal. An interesting aside about Camras: during World War II, his wire recorders were used by the military to train pilots. Battle sounds were recorded and equipment was developed to amplify the signal by thousands of watts to simulate the audio level of true battle. The most interesting application of his technology was deception, however. During the D-Day invasion, recordings were played at extremely high levels at

Figure 1-11

Camras receiving the National Medal of Technology from President Bush in 1990

various venues where the invasion of D-Day was *not* to take place, fooling the Germans into redirecting their forces. Because of the sensitive nature of his work, the public did not learn of Camras' contributions until long after the war had ended. Camras received more than 500 patents, largely in the field of electronic communications.

Early audiotape machines typically moved the tape along at a stately 15 inches per second. To meet the bandwidth requirements of the video signal, the tape had to be accelerated to somewhere between 300 and 400 inches per second—roughly 25 miles per hour. To put this into perspective, a tape that would accommodate an hour's worth of audio in those days would hold one minute of video—which did not take into account the length of the leader that had to be in place to enable the recorder to reach its ridiculously high tape transport speed. To hold 15 minutes of video, a reel of quarter-inch tape would have had to be 3 feet in diameter, which is not exactly portable. Put another way, a 1-hour show required 25 miles of tape!

To get around this problem, Camras invented the spinning record head. Instead of moving the tape rapidly past the recording head, he moved the tape slowly and rapidly spun the head. By attaching the head to a 20,000-RPM Hoover vacuum cleaner motor (stolen, by the way, from his wife's vacuum cleaner), he was able to use 2-inch tape and reduce the tape transport speed to 30 to 40 inches per second—a dramatic improvement.

The first video demonstrations were admirable, but rather funny. First of all, the resolution of the television was only 40 lines per inch compared to much higher resolution on modern systems. The images were so poor that audiences required a narrator to tell them what they were seeing on the screen.

Luckily, other advances followed. The original video systems rendered black-and-white images, but soon a color system was developed. It recorded 5 tracks (red, blue, green, synch, and audio) on half-inch tape and ran at the original speed of 360 inches per second. The system was a bit unwieldy. Because of the tremendous bandwidth that was necessary, it required one mile of color tape to capture a four-minute recording.

Obviously, mile-long tapes were unacceptable, especially if they only yielded four-minute programs. As a result, the Sarnoff/Ampex team reexamined the design of the recording mechanism. Three scientists—Charles Ginsburg, Alex Maxey, and Ray Dolby (later to be known for his work in audio)—redesigned the rotating record head, rotating it about 90 degrees so that the video signal was written on the tape in a zigzag design, as shown in Figure 1-12. This redesign, combined with *frequency modulation* (FM) instead of *amplitude modulation* (AM), allowed the team to reduce the tape speed to a remarkable 17½ inches per second. For comparison's sake, modern machines consume tape at about 2½ inches per second. This, by the way, is why the record head in your home VCR sits at a funny angle. (It's that big silver cylinder you see when you open the slot where the tape is inserted—see Figure 1-13.)

Figure 1-12
Zigzag writing method on modern video tape systems

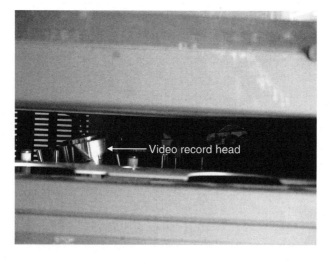

Figure 1-13
Video record head in a modern VCR— note the angle of the head

By 1956, Sarnoff and Ampex had created a commercially viable product. They demonstrated the "Mark IV" to 200 CBS affiliate managers in April of that year. When David Sarnoff walked out on the stage and stood next to his own prerecorded image playing on the television next to him (the audience thought they were watching him on a live broadcast), the room went berserk. In four days, Ampex took $5 million in video machine orders.

Modern Video Technology

Today's palm-size videotape recorders are a far cry from the washer-dryer-size Mark IV of 1956. But physical dimensions are only a piece of the video story.

The first video systems were analog and relied on a technique called *composite video*. In composite video, all of the signal components, such as the color, brightness, audio, synch, blanking, and so on, are combined into a single multiplexed signal. Because of the interleaved nature of this technique, a composite signal is complex and not particularly good. It suffers from impairments such as clarity loss between tape generations (in much the same way an analog audio signal suffers over distance) and color bleeding. Unfortunately, at the time video was developed, bandwidth was extremely expensive and the cost to transport five distinct high-bandwidth channels was inordinately high. Composite video, therefore, was a reasonable solution. Today, composite video is quite common because of technology improvements and the plummeting cost of electronics and bandwidth, and in fact, both *National Television System Committee* (NTSC) and *Phased Alternate Line* (PAL), the most commonly deployed television standards, are composite standards. These standards will be discussed later in the section "Destination Signal Issues."

As an alternative to composite video, *component video* soon emerged in which the signal components are transported separately, each in its own channel. This eliminated many of the impairments that plagued composite systems.

Several distinct color formats have emerged including RGB (for red, green, and blue, used in computer monitors); YUV (for luminance [Y], hue [U], and saturation [V]), often used in European PAL systems; YIQ (Y for luminance, and I and Q for color), typically used in NTSC systems; and a number of others. Luminance is analogous to the brightness of the video signal, while hue and saturation make up the chrominance (color) component.

All of these techniques accomplish the task of representing the RGB components needed to create a color video signal. In fact, they are mathematical permutations of one another.

One final observation is that RGB, YUV, YIQ, and the others are *signal formats*, which are different from *tape formats*, such as D1, D2, Betacam, VHS, S-VHS, DV, DV-CAM, and others. Signal formats describe the manner in which the information that represents the captured image is created and transported; tape formats define how the information is encoded on the storage medium. D1, for example, a component video standard, is used for very high-quality digital tape decks, whereas D2 and D3, both composite standards, provide medium- to high-quality image capture and storage.

Digital Video

Just as digital data transmission was viewed as a way to eliminate analog signal impairments, digital video formats were created to do the same for video. Professional formats such as D1, D2, Digital Betacam (the latter sometimes discounted because it incorporates a form of compression), and even High-8, DVCAM, and MiniDV virtually eliminate the problem of generational loss.

One downside is that even though these formats are digital, they still record their images sequentially on videotape. Today, video and computer technologies are married as nonlinear video systems for editing and management ease. Because the original video signal is digital, it can be moved from the digital tape captured by the camera to a large hard drive, where it can then be manipulated by video-editing software such as Adobe Premier or Apple's Final Cut Pro.

Today, the market demands an inexpensive, high-quality solution for the storage of video—CD-ROM and DVD are popular targets for such files.

The Video Process

A single CD-ROM holds approximately 650MB of data. The best numbers available today, including optimal sampling and compression rates, indicate that a VHS-quality recording requires roughly 5MB of storage per second of recorded movie. That's 7,200 seconds for a 2-hour movie, or roughly 30 CD-ROM disks. Finally, the maximum transfer rate across a typical bus in a PC is 420 Kbps, which is somewhat less than that required for a full-motion movie. Unless we want to abandon the CD in favor of a DVD, which has massive storage capacity, some form of compression must be invoked. Compression technologies will be discussed later in the section "Compressing Moving Images."

Video Summary

Let's review the factors and choices involved in creating video.

The actual image is captured by a camera that is either analog or digital (although for commercial applications, it is almost certainly digital). The resulting signal is then encoded to tape, after which it is transferred to CD-ROM or DVD or prepared for network streaming.

During the last decade, video has achieved a role of some significance in a wide variety of industries. It has found a home in medicine, travel, engineering, and education, and provides not only a medium for the presentation of information, but when combined with telecommunications technology, it also makes applications such as distance learning, video teleconferencing, and desktop video possible. The ability to digitize video signals has brought about a fundamental change in the way video is created, edited, transported, and stored.

Its emergence has also changed the players in the game. Once the exclusive domain of filmmakers and television studios, video is now fought over by creative companies and individuals who want to control its content, cable and telephone companies who want to control its delivery, and a powerful market that wants it to be ubiquitous, richly featured, and cheap. Regulators are in the mix as well, trying to make sense of a telecommunications industry that was once designed to transport voice but now carries a broad mix of fundamentally indistinguishable data types.

So far, we have discussed typical conferencing applications as well as the variety of conferencing media that exists. We talked about the history of video, the creation of video content, and the various ways it can be distributed. Before we introduce access and transport networking technologies, we must first discuss one video creation environment: the television studio. In this next section, we will describe the typical broadcast studio, how it works, how a presenter can take best advantage of the capabilities it offers, and what a presenter must do to prepare for the experience of conducting a session from a live studio environment. Although most videoconferences will not involve a commercial or corporate broadcast studio, some, such as training sessions, will. It is therefore important that we at least familiarize the reader with the studio environment. We will discuss the videoconferencing room later in the book.

The Studio   The first three things that a person realizes the first time he or she enters a television studio are that (1) it is awfully cold in here, (2) it is awfully gloomy in here, and (3) this place is a mess! The cold is a necessary evil because the lights generate an enormous amount of heat. Once they are turned on, the studio warms up quickly.

The second issue, the gloominess of the place, is simply because all surfaces are painted black or dark gray to reduce the possibility of stray light reflecting into the camera. Once the broadcast begins and the lights flare up, the gloominess goes away.

The final thing is the messiness of the place. There is no question that broadcast studios look like a riot just passed through them. There are cables, often several inches deep, running everywhere, equipment stuffed into every possible corner, alternate set dressings leaning against every wall, every kind of light imaginable hanging from the overhead lighting grid or perched atop light stands, and people running around with a terrible sense of purpose that is reminiscent of a *M*A*S*H* unit. Yet the viewer sees none of that. The viewer sees a calm, beautiful scene, a professional speaker who sounds terrific, and a show that always begins and ends right on time. What the viewer sees, as shown in Figure 1-14, is quite different from what the presenter sees (see Figure 1-15)!

Let's look at the layout of a typical studio, as shown in Figure 1-16. The presenter sits or stands at a credenza from which he or she can operate all of the equipment required to conduct the event. (A typical presenter's console is shown in Figure 1-17.) There is typically a

Figure 1-14
The viewer's perspective in a studio-based videoconference

Figure 1-15
The presenter's view is somewhat different from the participant's view!

Figure 1-16
Layout of a typical
video broadcast
studio

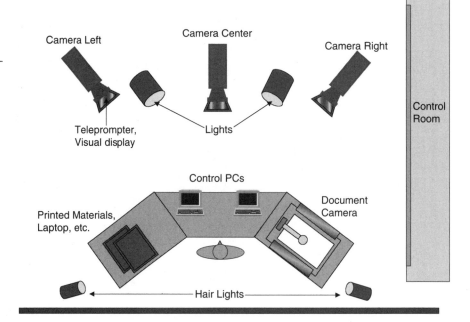

Camera Left

Camera Center

Camera Right

Teleprompter,
Visual display

Lights

Control
Room

Control PCs

Printed Materials,
Laptop, etc.

Document
Camera

Hair Lights

Figure 1-17
The presenter's
console, showing
(from left to right) a
document camera,
a prompter, and
PCs

document camera that enables the presenter to draw on a piece of paper and have the viewers see what is being drawn (it is nothing more than an additional camera, but instead of pointing at the presenter, it points downward at a flat illuminated surface); one or more PCs that allow the presenter to advance the slides if they are to be used; and space for written material, notes, and so on. The presenter may wear a device called an *ear prompter* (you can barely see it in my right ear in Figure 1-18), which enables the control room to communicate with the presenter. (A note to would-be presenters: If the director talks to you over the ear prompter during a broadcast, do *not* answer verbally. The audience cannot hear the director's voice so if you answer, the audience will think you're hearing voices in your head. The fact that you are doesn't matter.) Typically, the director will softly say things like "10 minutes to go" or "we have a question from Columbus," which you can acknowledge by slightly nodding your head.

The control room, as shown in Figure 1-19, is a soundproof room from which the director, sound engineer, and producer run the show. They control sound levels, camera settings, lighting levels, and the presenter. There may or may not be a person in the studio during the

Figure 1-18

Using an ear prompter to communicate with the control room

Figure 1-19
The complex view
in the control room

presentation. Occasionally, there are camera operators, but more commonly the cameras are locked down and do not require anyone to move or operate them. The control room, or in some cases, the presenter, can switch back and forth between different camera angles as required.

The presenter has a fairly limited view. Directly in front of the credenza are the cameras themselves and hanging below them are monitors that serve as teleprompters or as computer monitors that display the presenter's PowerPoint slides. The cameras are usually marked with numbers or letters. This is quite helpful since the director will occasionally whisper "camera two" into the presenter's ear, meaning "Look at camera two."

The presenter's view is rather harsh, as shown in Figure 1-20. In addition to the studio lights that illuminate the set (see Figure 1-21), there may be a ring of brilliant blue lights surrounding each camera. These are the lights that facilitate blue screen technology, which is widely used in television production today. It is most commonly seen

Figure 1-20
The harsh lighting
in the studio

Figure 1-21
The overhead
lighting grid in a
broadcast studio

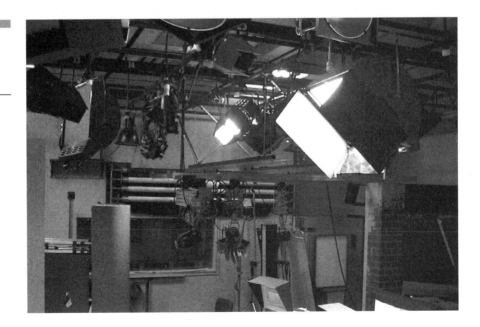

during the news. You may have noticed that when the weather person presents the weather with the map in the background, he or she does not actually look at the map. Instead, they look at the floor, where the monitor is located. (If you haven't noticed this before, pay attention the next time you watch the weather.) There is actually nothing on the wall; instead, the presenter's image is digitally overlaid with the weather map (or whatever background is required at the time). The technology is called blue screen or green screen because the wall behind the presenter is one of those colors, matching the lights that surround the camera. The digital camera and recording equipment are tuned so that the blue or green screen is not visible, making it possible to overlay an image digitally behind the presenter. Obviously, the presenter should not wear clothing of that color because as soon as the lights come on the clothing . . . disappears!

There is something fearsome about presenting on television for the first time. It is a different environment and requires a very different set of skills. A person who is accustomed to presenting in a classroom or conference room can be good on television, but a certain amount of preparation and seasoning must take place first. In a television studio, all kinds of unpredictable things can happen and the presenter must be prepared to deal with them because it is his or her show—all eyes are on the presenter. A few years ago I delivered a series of three-hour broadcasts to a client about optical-networking technologies. In one particular broadcast, I had arranged for a guest speaker to come into the studio following a break in the middle of the broadcast to discuss the merits of this particular company's products. The sound engineer had already threaded a wireless microphone under her blouse, and she was sitting quietly in the far corner of the studio. The director had already told her to be very quiet because her mike was on. I was wearing a wireless lavalier microphone as well as an ear prompter. The microphone was clipped to my shirt and the mike body, which was about the size of a pack of cigarettes, was in my right front pocket.

Suddenly, about 20 minutes before the scheduled break, the director's voice filled my ear through the prompter. "We are losing your microphone (the battery was dying). Karen (Karen could also hear him through her own prompter), I need you to crawl on your hands and knees behind the credenza and sit next to Steve. Your microphone is fine; as long as you are as close to him as possible, we'll be able to pick up his voice through your mike until the break. Whatever you do, don't allow your head to rise above the top of the credenza!"

You understand, of course, that I was lecturing to the television audience about some optical networking thing throughout this entire process, both ignoring what was being said and listening very carefully. I continued to talk right through Karen's migration to my side and was only aware of her presence because I could hear her. So, I continued to lecture.

The next unexpected thing was the hand that I suddenly felt in my pants pocket. What I learned later was that the sound engineer had the brilliant idea to sneak into the studio on his hands and knees, remove the mike body from my pocket, replace the batteries, put it *back* in my pocket, and crawl back out of the studio to the safety of the control room. The only flaw in this plan was that the sound engineer forgot to tell the floor director that he was going to do this, which meant that there was no one to warn me about the hand that was about to snake into my pocket.

The remarkable thing about this story is that the audience never knew. Karen came close to suffocating as she tried not to laugh as she watched my face from below the credenza. The only thing that saved me was that I have done several hundred video broadcasts over the years and know that anything can happen.

Preparing the Presenter   As a result of the growing corporate popularity of television education, conventional presenters are finding themselves thrust in front of commercial video cameras, an experience that many would forego in favor of a root canal if given a choice. Television is not a particularly easy medium to present with, but it doesn't have to be a grueling ordeal either. A well-prepared presenter will find that it isn't all that different from classroom or meeting room presentations, with a few notable exceptions. The following lists the key concerns a presenter should have:

■ *Rehearse, rehearse, rehearse. Then rehearse some more.*   When you feel as if you are truly comfortable and prepared for the show, rehearse again. There is no such thing as too much preparation, although heavy rehearsal the day before the broadcast can be overwhelming. Use the day before a broadcast to go over the logistics of the studio. Work with the director and studio staff to review the things that are important to them, such as hand signals, television methodology, and break logistics. The better you know their requirements and the closer you follow their instructions, the better your production will be.

■ *Watch others.*   Take time to watch professional broadcasters on television. Practiced lecturers and commercial broadcasters can lend tremendous insights into things such as mannerisms and actions that are (and are not) effective in a television medium. Watch what they do with their hands, listen to the cadence of their speech, and pay close attention to general body language. What kinds of things do they do that you find appealing? Conversely, what do they do that irritate you?

■ *Personal preshow care.*   Take care of yourself during the week preceding the broadcast. Televised instruction demands a deceptively large amount of emotional and physical energy so prepare your body for the ordeal. Eat light, healthy foods such as fish, chicken, lots of fruit, and vegetables. Minimize alcohol consumption (don't fall prey to the misconception that it will help you relax in the days before the broadcast). Get moderate exercise and lots of sleep. This is a good time to be selfish. Explain to those close to you and to bosses and peers that you won't be available to them the day prior to the telecast. Self-indulgence isn't a luxury at this stage of the game—it's a necessity.

■ *Trust the director.*   Put absolute faith and trust in the director. The director's job is to worry about the logistics of the television show so that you as the presenter (sometimes called "the talent" by the TV folks) don't have to. That leaves you alone to concern yourself with the material that you will present on the air. I was extremely lucky during my broadcast in that my director exuded professionalism. For the period leading up to the broadcast, I belonged to her and anyone who tried to get near me for anything that didn't have to do with the broadcast was in danger of losing ears and fingers. She took care of everything that had to do with the medium and in fact shooed me away from it wherever possible. Television production is fascinating, and there is a natural desire for most people to get wrapped up in it. She gave me a little time to be starry-eyed in the studio and once the fascination began to ebb, she prodded me back to work.

Don't be afraid to ask the director questions regarding the direction of the show. Remember, he or she probably won't know anything about the material that will be presented, but 99 percent of the time he or she doesn't have to. Occasionally, a situation will arise in which the nature of the material will dictate a particular direction that contradicts what the director wants to do. Don't be afraid to make suggestions, but if the director is adamant, that's okay. The director's job is to make you look good on screen. Your job is to present the material professionally. Period.

- *Clothing color.*   This is an important consideration during television instruction. Colors don't look the same on TV as they do live, and some are not recommended. White shirts are definitely *verboten*. White tends to flare on a video camera; light blue, dark colors, and tan are fine. Avoid patterned material at all costs—it creates a distracting moiré effect on camera.

  Similarly, single color suits in dark colors are great — no plaids, no herringbone, and no bright pinstripes. The simpler the material, the better it — and you — will look onscreen. Of course, avoid royal blues or bright greens if blue screen technology will be used during the broadcast. It is best to assume that it will be and dress accordingly. Be advised: if your clothing is not appropriate for the broadcast, the director will dress you in something that is.

- *Wear comfortable clothes and comfortable shoes.*   You will spend a considerable amount of time under hot lights so the lighter the clothing, the better. You may also be on your feet for extended periods of time so unless your particular production will show your feet, consider wearing sneakers.

- *Site coordinators are a necessity if you broadcast to multiple remote locations.*   They should receive formal training on their responsibilities and perform the following tasks. They ensure that textbooks and other materials arrive on time and that there are enough in each room for the number of registered participants. They set up the room and arrive early each day of the broadcast to ensure that the unit is on and that the audio and video signals are at their correct levels. They introduce themselves to the participants and give them a preclass overview that covers the delivery medium, the use of the telephone lines, what to do in case of trouble, how to ask questions, and so on. They make themselves available to conference participants during the day in case a problem crops up that needs their attention. In short, they represent another layer of responsibility that the presenter needn't worry about.

- *Participants in the studio.* There are two trains of thought on having a live audience in the studio. One says that participants in the studio lend comfort to the presenter since they provide eyeballs to focus on and an audience that can help you measure reaction to the material. The other thought is that the presence of people in the studio causes the presenter to have, in effect, two audiences. The natural tendency is for presenters to look at the people and ignore the camera. However, this could alienate remote participants.

  I personally find an empty studio to be easier. It takes some time to become comfortable with the impersonal nature of the camera, but that's what rehearsals are for. Presenters learn early on to use "faces in the crowd" to measure comprehension and control delivery timing. The television medium removes those visual cues, so a significant adjustment is often necessary. The presenter must resort to forced interaction—"Bob out in St. Louis, does your group have any questions? How about Dallas? Any questions there?"

- *Cover the logistics of the television medium before you begin.* Explain to participants what they can expect in terms of teaching technique. Cover the technology involved: how to call with a question and how to restore a lost signal (or who to call). Put contact/trouble numbers on the screen and ask participants to write them down. If you're lucky, one person at each location will be motivated to do so.

■ *Be yourself.*   Let's face it, standing in front of a television camera for a live broadcast is at best stressful and at worst terrifying. A strong mental resolve to keep things in perspective can help presenters get through it.

Try to remember that to the audience participants, you are similar (but not identical) to any other presenter that they have had in a live setting. The delivery medium is different, but to them, making a mistake on TV is no different than making a mistake in front of a live audience—and we've all done that. It's not the end of the world.

One key difference between a televised presenter and a live one is that for some unexplainable reason, television gives presenters a presence that is grander than life. Everything said to a group will be taken with a little more credence simply because it is delivered via television. For this reason alone, it is imperative that television presenters work hard to appear natural and comfortable on the air and to be willing to make mistakes—more importantly, to be willing to say, "I don't know, I'll get back to you on that one." The medium works both ways. It can remove credibility with just as much force as it can deliver it.

Don't be afraid to use notes or an Instructor Guide just because you are teaching on television. If that's the way you teach most effectively in a classroom environment, then by all means, teach that way on television. Again, this is not a measure of your on-screen presence as a commentator; it is just another presentation. If you regularly look down at notes when you speak before a live audience, don't be afraid to do it on TV. A word of advice: because of the exaggerated nature of television instruction, 5 seconds of reading notes on the desk feels like 50. Force yourself to take as much time as you need. Again, the best way to *look* comfortable is to *feel* comfortable, and the best way to do that is to minimize activities that take you out of your comfort zone.

■ *Use the "Larry King effect."*   This is the reticence that participants typically feel about calling a television presenter with a question. Many are reluctant to ask a live presenter a question, much less a TV personality.

To minimize this phenomenon, presenters can design their presentations in such a way that participants are ever so gently forced to interact. For example, remote sites can be asked to work together for a few minutes on a group project related to the subject matter. Then, the presenter can ask someone from each location to give feedback to the audience on their findings. This tends to make the speaker feel as if he or she is not alone and that he or she has a team behind them.

Other methods that encourage interaction include impromptu questions directed at particular sites or individuals at those sites. This can be a bit awkward, however, if the question requires some thought since the facilitator might be faced with managing dead air time—the first deadly sin of television— while an answer is being formulated. The presenter also runs the risk of alienating the audience by putting participants on the spot and potentially on the defensive. Play it by ear and take the time to know your audience prior to and during the broadcast. The better you know them, the better you will be able to assess the relative effectiveness of different techniques.

- *Variety is everything.* Even the best television shows and movies can become tiresome, and let's face it, folks, technical presentations ain't Bruce Willis. Hours in front of a television screen can result in eyestrain and a restless class, and the endless drone of a presenter's voice can mesmerize even the most dedicated participant into an unconscious stupor. To minimize this, break your presentation style up as much as possible. Intersperse the sight of your face on the screen with short video segments, and use a video projector or document camera to show illustrations or draw pictures. Give the audience frequent breaks —at least five minutes each hour, and longer breaks at mid-morning and mid-afternoon—and allow for off-air time, when you give them off-camera exercises to do.

- *Be creative.* Have sites prepare exercise solutions and fax them to you in the studio, where you can put them up on the screen and share them with others. Encourage sites to talk to each other and encourage participants to talk to you. Remember, variety increases your chances of having an attentive class; avoid the "talking head" syndrome.

■ *If you plan to project graphics over the network using a document camera or scan converter (which concerts the output of a PC so that it can be shown on a TV), follow these guidelines.* Minimum type size should be 36 point and all graphics should be printed on matte finish paper—no gloss. Line art should be black and white, uncluttered, and simple. The typical participant will be watching the show on a 21-inch television set from a distance of at least 10 feet. Detailed pictures are a wasted effort. Instead, have multiple copies of each graphic and use a heavy, black felt-tip pen to enhance the drawing by hand as you speak to the illustration. This is a very effective technique for gaining the participant's attention. One last tip: check with the director to see whether graphics should be printed vertically or in landscape orientation. Requirements differ between devices. It is highly likely that studio personnel will have a PowerPoint template for you to use that is "TV safe"—meaning that it shows the areas on the slide where content must be placed to appear properly on TV.

Television can be an extremely effective delivery medium if the presenter is adequately prepared for the "culture shock" that usually accompanies a broadcast. Proper support from the television technical staff, absolute familiarity with the material, and comfort with the technology in the studio are requirements for a successful broadcast.

Other Considerations In addition to broadcast video, a number of other delivery media are commonly used as an alternative to a live presentation. These may include conference calls, posting the materials on a web site, or the distribution of a PowerPoint presentation via e-mail. Unfortunately, these techniques often fall short because of a few minor issues that can be easily remedied:

■ Do not use these techniques for the presentation of highly technical or overly complex material. This type of information benefits greatly from the use of visual presentation components combined with an instructor and is therefore not effectively presented using virtual presentation techniques.

- To the extent possible, limit the number of participants if the selected medium is a webcast or telephone conference call. If the audience becomes too large, effective question-and-answer interaction is hampered.

- Prepare handouts and make sure that they are in the hands of all attendees at least a day before the actual presentation. This will ensure that attendees have an ample opportunity to review the content and prepare questions in advance.

- Handouts should be carefully sequenced and should answer the following questions:

  - What is the purpose of the presentation?
  - How long will it last?
  - Who is the intended audience?
  - What is the mechanism for follow up? Should questions arise after the presentation?

- Handouts should be designed so that major discussion points are presented in a logically isolated fashion. In other words, do not use slides that present multiple ideas or concepts. It is better to use more simple slides than fewer complex ones.

- Avoid overly complicated diagrams that are difficult to understand. Remember that the audience does not have the benefit of an instructor at the front of the room holding their hands as they go through the material.

- Make it a point to build natural break points into the presentation to allow attendees to ask questions. In fact, it isn't a bad idea to seed the audience with "plants" who ask prepared questions to get things flowing.

- Do not run over the advertised time interval. Make it clear that the presentation is over at a defined time, but that you are willing to stay on the line to discuss anything that the audience wants to discuss.

- Coach moderators to speak clearly and enunciate. Where possible, the moderator should not use a conference call to originate the discussion; a telephone handset or headphone yields much higher-quality voice.

- If collateral materials are to be made available (highly recommended), include a clear description of what they are and where they can be found. For example, if information is to be included on a web site, be sure to include the URL of the site in the presentation materials and discuss what the students will find there (and the value of the information to the participants).

- If possible, offer the presentation more than once.

- If the presentation will be delivered to an international or multinational audience, be sensitive to that fact. Use country-specific examples and units of measure, and pay attention to national holidays.

The End-User Environment

In this section we discuss the components found in a typical end-user environment including computers and LANs. We begin with an examination of the computer. In one way or another, the computer is the ultimate technology device; it appears in one form or another in every device used by a customer to access the network, including videoconference units.

The Computer

For all its complexity, the typical computer only has a small number of components, as shown in Figure 1-22: the *central processing unit* (CPU), main memory, secondary memory, *input/output* (I/O) devices, and a parallel bus that ties all the components together. It also has two types of software that make the computer useful to a human. The first is application software such as word processors, spreadsheet applications, presentation software, and MP3 encoders. The second is the operating system that manages the goings on within the computer including hardware component inventory and file location. In a sense, it is the executive assistant to the computer itself;

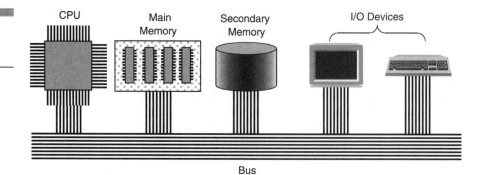

Figure 1-22
Computer components

some mainframe manufacturers refer to their operating system as the EXEC.

The concept of building modular computers came about in the 1940s when Hungarian-born mathematician John Von Neumann applied the work he had done in logic and game theory to the challenge of building large electronic computers (see Figure 1-23). As one of the primary contributors to the design of the *Electronic Numerical Integrator and Computer* (ENIAC), Von Neumann introduced the concept of stored program control and modular computing, the design under which all modern computers are built today.

A personal computer's internals are shown in Figure 1-24.

The CPU   The CPU is the brain of the computer. Its job is to receive data input from the I/O devices (the keyboard, mouse, modem, and so on), manipulate the data in some way based on a set of commands from a resident application, and package the newly gerrymandered data for presentation to a user at another I/O device (monitor). The CPU has its own set of subcomponents. These include a clock, an *arithmetic-logic unit* (ALU), and registers. The clock is the device that provides synchronization and timing to all devices in the computer and is characterized by the number of clock cycles per second it is capable of generating. These cycles are called *hertz* (Hz); modern systems today operate at 850 to 1,000 *megahertz* (MHz). The faster the clock, the faster the machine can perform computing operations. Obviously, for graphics or multimedia-intensive applications, the faster the processor, the better the performance of the applica-

Figure 1-23
*A section of the
original ENIAC
machine*

tion because a tremendous amount of processing takes place during the rendering of image files.

The ALU is the specialized silicon intelligence in the CPU that performs the mathematical permutations that make the CPU useful. All functions performed by a computer—word processing, spreadsheets, multimedia, and videoconferencing—are viewed by the computer as mathematical functions and are therefore executed as such. It is the ALU's job to carry out these mathematical permutations.

Registers are nothing more than very fast memory located close to the ALU for rapid I/O functions during execution cycles.

Figure 1-24
PC internals
showing major
components

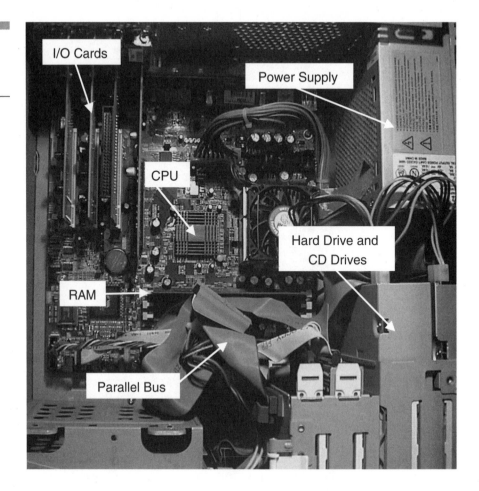

Main Memory Main memory, sometimes called *random access memory* (RAM), is another measure of the "goodness" of a computer today. RAM is the segment of memory in a computer used for execution space and as a place to store operating system, data, and application files that are in current use. RAM is silicon-based, solid-state memory and is extremely fast in terms of access speed. It is, however, volatile. When the PC is turned off, or power is lost, whatever information is stored in RAM disappears. This is the reason why basic computer-skills courses tell students to save often. When a computer user is writing a document in a word processor, populating a spreadsheet, or manipulating a digital photograph, the file lives in RAM until the person saves, at which time it is written to main memory, which is nonvolatile as we'll see in a moment. Modern systems typically have a minimum of 128MB of RAM.

Secondary Memory Secondary memory has become a very popular line of business in the evolving PC market. It provides a mechanism for the long-term storage of data files and is nonvolatile—when the power goes away, the information it stores does not. Secondary memory tends to be a much slower medium in terms of access time than main memory because it is usually mechanical in nature, whereas main memory is solid state. (Consider, for example, the hard drive shown in Figure 1-25.) That's why the disk platters are exposed. Notice that the drive comprises three platters and an armature where read/write heads similar to those used in cassette decks of yore are mounted. Figure 1-26 is a schematic diagram that illustrates how hard drives actually work. The platters, typically made of aluminum and cast to extremely exacting standards, are coated with iron oxide that is identical to what is found on recording tape. Adjacent to the platters, which are mounted on a spindle attached to a high-speed motor, is an armature where a stack of read-write heads that write information to the disk surfaces and read information from the disk surfaces are mounted. Remember when we were discussing the importance of saving often when working on a computer? The SAVE process reads information stored temporarily in RAM, transports the information to the disk controller, which in turn, under the guidance of the operating system, transmits the information to the write heads, which in turn writes the bits to the disk surface by magnetizing the iron oxide surface. Of course, once the bits have been written to the disk, it's important to keep track of where the information is stored on the disk. This is a function of the operating system. Have you ever inserted a floppy into a drive or (far less fun) a new hard drive into a computer as I recently had to do, and had the computer ask you if you're sure you want it to format the drive? In order to keep track of where files are stored on a disk, whether it is on a hard drive, a CD, a Zip drive, or a floppy, the operating system marks the disk with a collection of road markers that help it find the "beginning" of the disk so that it can use that as a reference. Obviously, there is no beginning on a circle by design.

Figure 1-25
Close-up of a hard
drive, showing
platters and read-
write head
armature

Read-Write Heads

Platters

Figure 1-26
Schematic diagram
of a hard drive. As
the platters spin,
the read-write
heads access the
magnetized
surfaces of each
platter, writing
information to
and reading
information from
them.

By creating an arbitrary start point, however, the operating system can then find the files it stores on the drive.

When the operating system formats a disk, it logically divides the disk into what are called *cylinders*, *tracks*, and *sectors*, as shown in Figure 1-27. It uses those delimiters as a way to store and recall files on a drive using three dimensions. When the operating system writes the file to the disk surface, there is every possibility (in fact, it is likely) that the file will not be written to the surface in a contiguous stream of bits. Instead, the file may be broken into pieces, each of which may be written on a different part of the disk array. The pieces are linked together by the operating system using a series of pointers, which tell the operating system where to go to get the next piece of the file when the time comes to retrieve it. The pointer, of course, is a cylinder, track, and sector indication.

Cylinders, tracks, and sectors are relatively easy to understand. A track is a single writeable path on one platter of the disk array. A cylinder is a stack of tracks on multiple platters. A sector is a pie slice of the disk array. With a little imagination, it is easy to see how a file can be stored or located on a disk by tracking those three indicators.

It should also be easy to understand how applications like Norton Utilities work. When you direct your computer to erase a file, it doesn't really erase it; it just removes the pointers so that it is no longer a registered file and can no longer be found on the hard drive. File restore utilities remember where the pointers were when you told

Figure 1-27

Cylinders, tracks, and sectors on a typical disk

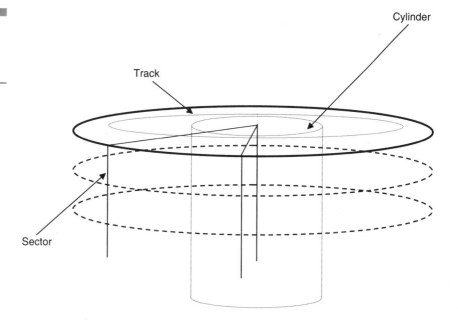

the computer to erase the file. By doing so, the utility has a trivial task: restore the pointers. As long as there hasn't been too much disk activity since the deletion and the file hasn't been overwritten, it can be recovered.

Another useful tool is the disk optimization utility. It keeps track of the files and applications that the user accesses most often while at the same time keeping track of the degree of fragmentation that the files stored on the disk are experiencing. Think about it: when disks start to get full, there is less of a chance that the operating system will find a single contiguous piece of writeable disk space and will therefore have to fragment the file before writing it to the disk. Disk optimization utilities perform two tasks. First, they rearrange files on the disk surface so that those accessed most frequently are written to new locations close to the spindle, which spins faster than the outer edge of the disk and is therefore accessible more quickly. Second, they rearrange the file segments and attempt to reassemble files or at least move them closer to each other so that they can be more efficiently managed.

Okay, enough about hard drive anatomy. Other forms of secondary memory include those mentioned earlier—writeable CDs, Zip disks, floppies, and nonvolatile memory arrays. As I mentioned, this has become a big business because these products make it relatively easy for users to back up files and keep their systems running efficiently.

Input/Output (I/O) Devices   I/O devices are those that provide an interface between the user and the computer and include mice, keyboards, monitors, printers, scanners, modems, speakers, and any other devices that take data in or spit data out of the computer.

The Bus   A cable known as a *parallel bus* interconnects the components of the computer. It is called "parallel" because the bits that make up one or more 8-bit bytes travel down the bus beside each other on individual conductors, rather than one after the next, as occurs in a serial cable on a single conductor. Both are shown schematically in Figure 1-28. The advantage of a parallel bus is speed. By pumping multiple bits into a device in the computer simultaneously, the device, such as a CPU, can process them faster. Obviously, the more leads there are in the bus, the more bits can be

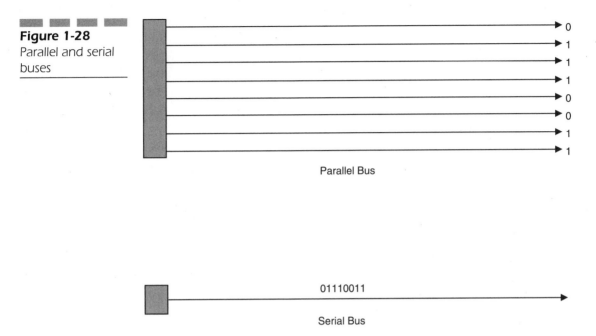

Figure 1-28

Parallel and serial buses

transported. It should come as no surprise, then, that another differentiator of computers today is the width of the bus. A 32-bit bus is 4 times faster than an 8-bit bus. As long as the internal device to which the bus is transporting data has as many I/O leads as the bus, it can handle the higher volume of traffic. The parallel bus does not have to be a flexible gray cable. For those devices that are physically attached to the main printed circuit board (often called the *motherboard*), the parallel bus is extended as wire leads that are etched into the surface of the motherboard. Devices such as I/O cards attach to the board via edge connectors.

There are many different types of computers, including specialized supercomputers, mainframes, minicomputers, personal computers, *personal digital assistants* (PDAs), and specialty servers designed to perform a single set of related tasks. For example, a streaming media server has a very fast processor, high-volume I/O ports, and a tremendous amount of memory so that it can handle the high-bandwidth requests for multimedia content that it will receive.

The real evolution, of course, came with the birth of the personal computer. Thanks to Bill Gates and his concept of a simple operating system (DOS) and Steve Jobs with his vision of "computing for the masses," truly ubiquitous computing has become a reality. From the perspective of the individual user, this evolution was unparalleled. The revolution began in January 1975 with the announcement of the MITS Altair (see Figure 1-29). Built by *Micro Instrumentation and Telemetry Systems* (MITS) in Albuquerque, New Mexico, the Altair was designed around the Intel 8080 microprocessor and a specially designed 100-pin connector. The machine ran a BASIC operating system developed by Bill Gates and Paul Allen; in effect, MITS was Microsoft's first customer.

The Altair was really a hobbyist's machine, but soon the market shifted to a small business focus. Machines like the Apple, Osborne, and Kaypro were smaller and offered integrated keyboards and video displays. In 1981, of course, IBM entered the market with the DOS-based personal PC, and soon the world was a very different place. Apple followed with the Macintosh and the first commercially available *graphical user interface* (GUI), and the rest, as they say, is history. Soon corporations embraced the PC. "A chicken in every pot,

Figure 1-29
The ALTAIR 8800 computer. Photo courtesy Jim Willing, The Computer Garage.

a computer on every desk" seemed to be the rallying cry for IT departments everywhere.

There was a downside to this evolution, of course. The arrival of the PC heralded the arrival of a new era in computing that allowed individuals to have their own applications, file structures, and data. The good news was that each person now controlled his or her own individual computer resources; the bad news was that each person now controlled his or her own individual computer resources. Suddenly, the control was gone. Instead of having a copy of the database, there were now as many copies as there were users. This led to huge problems. Furthermore, PC proliferation led to another challenge: connectivity, or lack of it. Whereas before every user had electronic access to every other user via the mainframe or minicomputer-based network that hooked everyone together, the PC did not have that advantage. Furthermore, PCs eliminated the efficiency with which expensive network resources, such as printers, could be shared. Some of you may remember the days when being the person in the office with the laser printer attached to your machine was akin to approaching a state of nirvana. You were able to do your work and print anytime you wanted—except, of course, for the disruption caused by all of those people standing behind you with floppy disks in their hands promising you that "It's only one page—it'll just take a second." Thus, the term *sneakernet* was born. In order to print something, you had to put the document on a diskette (probably a 5¼-inch floppy, back when floppies really were floppy), walk over to the machine with the directly attached printer, and beg and wheedle for permission to print the file—not the most efficient technique for sharing a printer. Something else was needed. That *something* was the LAN.

LAN Basics

A LAN is exactly what the name implies: a physically small network, typically characterized by high-speed transport, low error rates, and private ownership, and that serves the data transport needs of a geographically small community of users. Most LANs provide connectivity, resource sharing, and transport services within a single building, although they can operate within the confines of multiple buildings on a campus.

When LANs were first created, the idea was to design a network option that would provide a low-cost solution for transport. Up until their arrival, the only transport option available was a dedicated private line from the telephone company or X.25 packet switching. Both were expensive and X.25 was less reliable than was desired. Furthermore, the devices being connected together were relatively low-cost devices; it simply didn't make sense to interconnect them with expensive network resources. That would defeat the purpose of the LAN concept.

All LANs, regardless of the access mechanism, share certain characteristics. All rely on some form of transmission medium that is shared among all the users on the LAN, all use some kind of interrupt and contention protocol to ensure that all devices get an equal opportunity to use the shared medium, and all have some kind of software called a *network operating system* (NOS) that controls the environment.

All LANs have the same basic components, as shown in Figure 1-30. A collection of devices such as PCs, servers, and printers serves as the interface between the user and the shared medium. Each of

Figure 1-30
Typical LAN
components

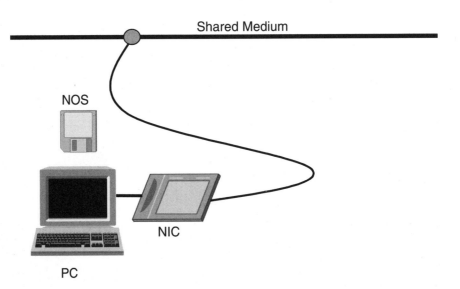

these machines hosts a device called a *network interface card* (NIC), which provides the physical connectivity between the user device and the shared medium. The NIC is either installed inside the system or less commonly as an external device. In laptop machines, the NIC is a PC card that plugs into a slot in the machine.

The NIC device implements the access protocol that devices wanting to access the shared medium use on their particular LAN. These access schemes will be discussed shortly. The NIC also provides the connectivity required for attaching a user device to the shared network.

Topologically, LANs differ greatly. The earliest LANs used a bus architecture, which is shown in Figure 1-31. These were called bus-based LANs because they were a long run of twisted-pair wire or coaxial cable to which stations were periodically attached. Attachment was easy; in fact, early coax systems relied on a device called a *vampire tap*, which poked a hole in the insulation surrounding the center conductor in order to suck the digital blood from the shared medium. Later designs such as IBM's Token Ring used a contiguous ring architecture such as that shown in Figure 1-32. Both architectures have their advantages and will be discussed in more detail later in this chapter.

Later designs combined the best of both topologies to create star-wired LANs (see Figure 1-33).

LAN Access Schemes

LAN have traditionally fallen into two primary categories characterized by the manner in which they access the shared transmission medium (shared among all the devices on the LAN). The first and most common category is called *contention*, and the second group is called *distributed polling*. I tend to refer to contention-based LANs

Figure 1-31
A bus-based LAN

Figure 1-32
A ring-based LAN

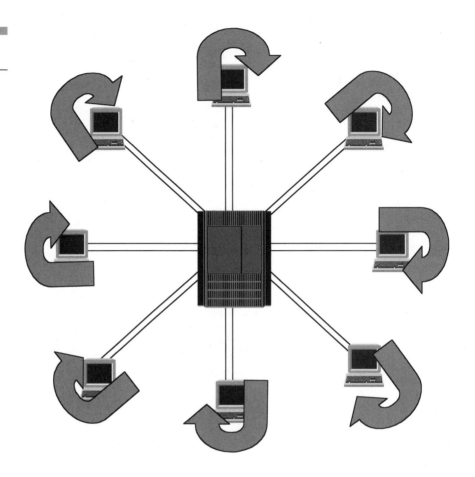

as the Berkeley Method, while I view distributed polling LANs as the Harvard Method. I'll explain in a moment.

Contention-Based LANs   Perhaps the best-known contention-based medium access scheme is the Ethernet, a product developed by 3Com founder and Xerox PARC veteran Bob Metcalfe. In contention-based LANs, devices attached to the network vie for access using the technological equivalent of gladiatorial combat. "If it feels good, do it" is a good way to describe the manner in which they share access (hence, the Berkeley Method). If a station wants to transmit, it simply does so, knowing that the possibility exists that the transmitted signal may collide with the signal generated by another station that transmits at the same time. Even though the transmissions

Figure 1-33
A star-wired LAN

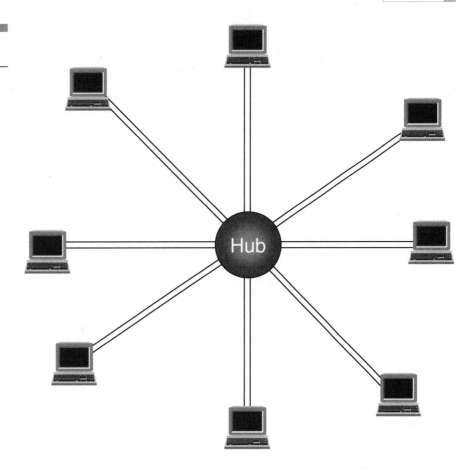

are electrical and are occurring on a LAN, there is still some delay between the time that both stations transmit and the time that they realize that someone else has transmitted. This realization is called a *collision* and it results in the destruction of both transmitted messages, as shown in Figure 1-34. In the event that a collision occurs as the result of simultaneous transmission, both stations back off by immediately stopping their transmissions, wait a random amount of time, and try again.

Ultimately, each station will get a turn to transmit, although how long the station will have to wait is based on how busy the LAN is. Contention-based systems are characterized by what is known as *unbounded delay* because there is no upward limit on how much delay a station can incur as it waits to use the shared medium. As the

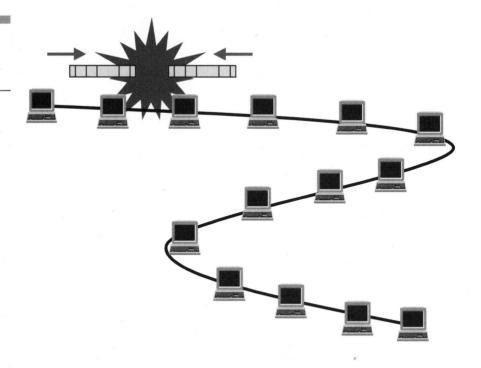

LAN becomes busier and traffic increases, the number of stations vying for access to the shared medium—which only permits a single station at a time to use it, by the way—also goes up, which naturally results in more collisions. Collisions translate into wasted bandwidth, so LANs do everything they can to avoid them. We will discuss techniques for this in the contention world a bit later in this chapter.

The protocol that contention-based LANs employ is called *Carrier Sense Multiple Access with Collision Detection* (CSMA/CD). In CSMA/CD, a station observes the following guidelines when attempting to use the shared network. First, it listens to the shared medium to determine whether it is in use or not—that's the Carrier Sense part of the name. If the LAN is available (not in use), it begins to transmit, but continues to listen while it is transmitting, knowing that another station could also choose to transmit at the same time —that's the Multiple Access part. In the event that a collision is detected, usually indicated by a dramatic increase in the signal power measured on the shared LAN, both stations back off and try again—that's the Collision Detection part.

Ethernet is the most common example of a CSMA/CD LAN. Originally released as a 10 Mbps product based on the IEEE standard 802.3, Ethernet rapidly became the most widely deployed LAN technology in the world. As bandwidth-hungry applications such as e-commerce, *Enterprise Resource Planning* (ERP), and web access evolved, transport technologies advanced, and bandwidth availability (and capability) grew, 10 Mbps Ethernet began to show its age. Today, new versions of Ethernet have emerged that offer 100 Mbps (Fast Ethernet) and 1,000 Mbps (Gigabit Ethernet) transport, with plans afoot for even faster versions. Furthermore, in keeping with the demands being placed on LANs by convergence, standards are evolving for LAN-based voice transport that guarantee QoS for mixed traffic types.

Gigabit Ethernet is still in a somewhat nascent stage, but most people believe that it will experience a high uptake rate as its popularity climbs. Dataquest predicts that Gigabit Ethernet sales will grow to $2.5 billion by 2002; this is a reasonable number, considering that 2 million ports were sold in 1999 with expectations of hitting a total installed base of 18 million by 2002. Emerging applications certainly make the case for Gigabit Ethernet's bandwidth capability. LAN telephony, server interconnection, and video to the desktop all demand low-latency solutions, and Gigabit Ethernet may be positioned to provide it. A number of vendors have entered the marketplace including Alcatel, Lucent Technologies, Nortel Networks, and Cisco Systems. It should also be noted that a new Ethernet standard, 802.3az, has emerged, which promises to deliver 10 Gbps service. This standard, designed primarily for the burgeoning metropolitan networking marketplace, may lend itself to video applications because of the bandwidth delivery that it makes possible.

The other aspect of the LAN environment that began to show weaknesses was the overall topology of the network itself. LANs are broadcast environments, which means that when a station transmits, every station on the LAN segment hears the message (see Figure 1-35). Although this is a simple implementation scheme, it is also wasteful of bandwidth because stations hear broadcasts that they have no reason to hear. In response to this, a technological evolution occurred. It was obvious to LAN implementers that the traffic on most LANs was somewhat domain oriented—that is, it tended to cluster into communities of interest based on the work groups using the LAN. For example, if employees in Sales shared a LAN with Shipping and Order Processing, three discernible traffic groupings would emerge according to what network architects call the *80:20 Rule*. The 80:20 Rule simply states that 80 percent of the traffic that originates in a particular work group tends to stay in that work group, an observation that makes network design distinctly simpler. If the traffic naturally tends to segregate itself into groupings, then the topology of the network could change to reflect those groupings. Thus, the bridge was born.

Bridges are devices with two responsibilities. They filter traffic that does not have to propagate in the forward direction and forward traffic that does. For example, if the network described previously were to have a bridge inserted in it (see Figure 1-36), all of the employees in each of the three work groups would share a LAN segment, and each segment would be attached to a port on the bridge. When an employee in Sales transmits a message to another employee in Sales, the bridge is intelligent enough to know that the traffic does not have to be forwarded to the other ports. Similarly, if the Sales employee sends a message to someone in Shipping, the bridge recognizes that the sender and receiver are on different segments and thus forwards the message to the appropriate port, using address information in a table that it maintains (the filter/forward database).

Following close on the heels of bridging is a relatively new technique called *LAN switching*. LAN switching could be described as "bridging on steroids." In LAN switching, the filter/forward database

Figure 1-35
When one station transmits, all stations hear the message. This can result in a significant waste of bandwidth.

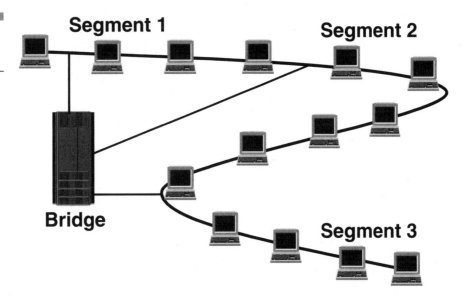

is distributed—that is, a copy of it exists at each port, which implies that different ports can make simultaneous traffic-handling decisions. This enables the LAN switch to implement full-duplex transmission, reduce overall throughput delay, and, in some cases, implement per-port rate adjustments. The first 10 Mbps Ethernet LAN switches emerged in 1993 and were followed closely by Fast Ethernet (100 Mbps) versions in 1995 and Gigabit Ethernet (1000 Mbps) switches in 1997. Fast Ethernet immediately stepped up to the marketplace bandwidth challenge and was quickly accepted as the next generation of Ethernet. LAN switching also helped to propagate the topology called *star wiring*. In a star-wired LAN, all stations are connected by wire runs back to the LAN switch or a hub that sits in the geographical center of the network, as shown in Figure 1-37. Any access scheme (contention based or distributed polling) can be implemented over this topology because it defines a wiring plan, not a functional design. Because all stations in the LAN are connected back to a center point, management, troubleshooting, and administration of the network are simplified.

Contention-based LANs are the most commonly deployed LAN topologies. Distributed polling environments, however, do have their place.

Figure 1-37
LAN switching

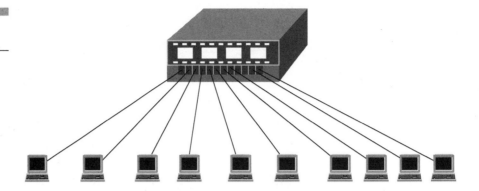

Distributed Polling LANs In addition to the gladiatorial combat approach to sharing access to a transmission facility, there is a more civilized technique known as distributed polling, or as it is more commonly known, *token passing*. IBM's token-passing ring is perhaps the best known of these products, followed closely by *Fiber Distributed Data Interface* (FDDI), a 100 Mbps version occasionally seen in campus and *metropolitan area networks* (MANs) (although the sun seems to be setting on FDDI).

In token-passing LANs, stations take turns with the shared medium, passing the right to use it from station to station by handing off a token that gives the bearer the one-time right to transmit while all other stations remain quiescent—thus, the Harvard Method. This is a much fairer way to share access to the transmission medium than CSMA/CD because although each station has to wait for its turn, it is absolutely guaranteed that it will get that turn. These systems are therefore characterized by bounded delay because there is a maximum amount of time that any station will ever have to wait for the token.

Token-passing rings work as shown in Figure 1-38. When a station wants to transmit a file to another station on the LAN, it must first wait for the token—a small and unique piece of code that must be held by a station to validate the frame of data that is created and transmitted. Let's assume for a moment that a station has secured the token because a prior station has released it. The station places the token in the appropriate field of the frame it builds (actually, the *media access control* (MAC) scheme, which is implemented on the

Figure 1-38
A token-passing,
distributed polling
LAN

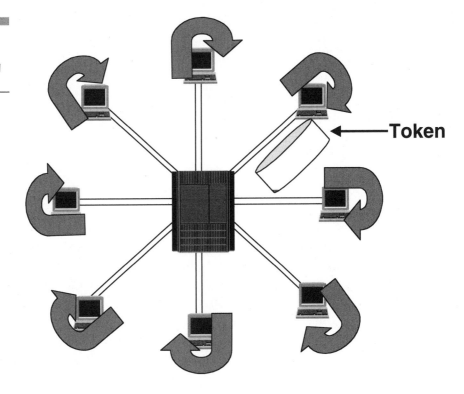

Token

NIC that builds the frame), adds the data and address, and transmits the frame to the next station on the ring. The next station, which also has a frame it wants to send, receives the frame, notes that it is not the intended recipient, and also notes that the token is busy. It does not transmit, but passes the frame of data from the first station onto the next station instead. This process continues, station by station, until the frame arrives at the intended recipient on the ring. The recipient validates that it is the intended recipient at which time it makes a copy of the received frame, sets a bit in the frame to indicate that it has been successfully received, leaves the token set as busy, and transmits the frame onto the next station on the ring. Because the token is still shown as busy, no other station can transmit. Ultimately, the frame returns to the originator at which time it is recognized as having been received correctly. The station therefore removes the frame from the ring, frees the token, and passes it onto the next station (it is not allowed to send again just because it is in possession of a free token).

This is where the overall fairness scheme of this technique shines through. The very next station to receive a free token is the station that first indicated a need for it. It will transmit its traffic after which it will pass the token onto the next station on the ring, followed by the next station, and so on.

This technique works very well in situations where high traffic congestion on the LAN is the norm. Stations will always have to wait for what is called *maximum token rotation time*—the amount of time it takes for the token to be passed completely around the ring—but they will *always* get a turn. Thus, for high-congestion situations, a token-passing environment may be better.

Traditional Token Ring LANs operate at two speeds—4 and 16 Mbps. Like Ethernet, these speeds were fine for the limited requirements of text-based LAN traffic that was characteristic of early LAN deployments. However, as demand for bandwidth climbed, the need to eliminate the bottleneck in the Token Ring domain emerged and fast Token Ring was born. In 1998, the IEEE 802.5 committee (the oversight committee for Token Ring technology) announced draft standards for 100 Mbps high-speed Token Ring (HSTR 802.5t). A number of vendors stepped up to the challenge and began to produce high-speed Token Ring equipment, including Madge Networks and IBM.

Gigabit Token Ring is on the horizon as draft standard 802.5v, and although it may become a full-fledged product, many believe that it may never reach commercial status because of competition from the much less expensive Gigabit Ethernet. In fact, recent discussions seem to indicate that the sun is setting on Token Ring in general; only time will tell. There is no question that Ethernet has the bulk of the marketplace. It would take a cataclysmic market event to change that, particularly given the success of Gigabit and 10 Gigabit Ethernet.

Logical LAN Design

One other topic that should be covered before we conclude our discussion of LANs is logical design. Two designs have emerged over the years for LAN data management. The first is called *peer to peer*. In a peer-to-peer LAN, shown in Figure 1-39, all stations on the network operate at the same protocol layer, and all have equal access at any time to the shared medium and other resources on the network. They do not have to wait for any kind of permission to transmit; they simply transmit. Traditional CSMA/CD is an example of this model. It is simple, easy to implement, and does not require a complex operating system to operate. It does, however, result in a free-for-all approach to networking, and in large networks, this can result in security and performance problems.

The alternative and far more commonly seen technique is called a *client-server LAN*. In a client-server LAN, all data and application resources are archived on a designated server that is attached to the LAN and accessible by all stations (user PCs) with the appropriate permissions, as illustrated in Figure 1-40. Because the server houses all of the data and application resources, client PCs do not have to be particularly robust. When a user wants to execute a program such as a word processor, he or she goes through the same keystrokes he or she would on a stand-alone PC. In a client-server environment, however, the application actually executes on the server, giving the user the appearance of local execution. Data files modified by the user are also stored on the server, resulting in a significant improvement in data management, cost control, security, and software harmonization compared to the peer-to-peer design. This also means that client devices can be relatively inexpensive because they need very little in the way of onboard computing resources. The server, on the other hand, is really a PC with additional disk, memory, and processor

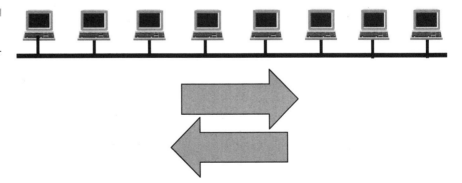

Figure 1-39
A peer-to-peer LAN

Figure 1-40
A client-server LAN

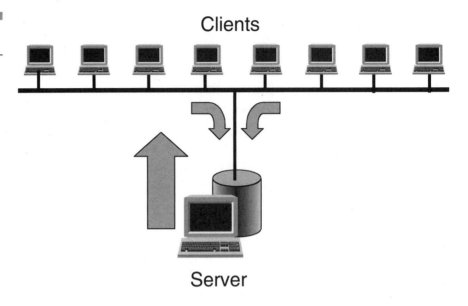

Clients

Server

capacity so that it can handle the requests it receives from all the users who depend on it. Needless to say, client-server architectures are more common today than peer-to-peer architectures in corporate environments.

Deployment

So when should Ethernet be used as opposed to Token Ring? Both have their advantages and disadvantages, both have solid industry support, and both are manufactured by a number of respectable, well-known players. CSMA/CD (for all intents and purposes, Ethernet) is far and away the most widely deployed LAN technology because it is simple, inexpensive, and capable of offering very high bandwidth to any marketplace, including residential.

Most businesses use Ethernet today because most businesses have normal traffic flows—office automation traffic and the like. For businesses that experience constant, bandwidth-intensive traffic such as that found in engineering firms, architectural enterprises, or businesses with other graphics-heavy traffic, Token Ring may be a better choice, although there are those who will argue against this statement. Businesses that already have a large installed base of IBM hardware may also be good candidates for Token Ring because it integrates well (for obvious reasons) into IBM environments. Even still, Ethernet is ruling the roost.

LANs and Videoconferencing

Because of the convergence of services that is underway in many corporations and the evolution of capability in enterprise networking (Gigabit Ethernet, for example), many IT managers are looking to converge enterprise traffic onto a common network infrastructure, specifically the corporate LAN. Because there is so much bandwidth available, they reason that all data, voice, and video traffic should be transportable across a single high-bandwidth network. This would reduce the overall cost of IT because the number of managed networks would be significantly reduced.

Unfortunately, there is a flaw in this logic. The data traffic that typically flows across a LAN and for which LANs were designed, is highly bursty, which means that it does not behave according to any predictable models. Of course, IT personnel know when the peaks occur, but they have a difficult time predicting the behavior of data traffic on a minute-by-minute basis. The other problem is that the data traffic that traverses most LANs is not terribly sensitive to delay because it comprises e-mail messages and other corporate traffic. It does not typically contain voice, video, or other traffic types that can be adversely affected by the variable delay that is an inherent characteristic of LANs.

When IT managers attempt to mix traffic with varying QoS demands on a common network backbone, problems can occur if the backbone lacks sufficient bandwidth to handle simultaneous peaking that will inevitably occur and if the backbone lacks a QoS policy enforcement mechanism. QoS is an issue in local, metro, and wide area transport networks. In the following paragraphs, we will discuss how QoS works across the network.

QoS begins at the customer location and is ideally maintained through the wide area to the receiving user device, covering the entire end-to-end connection. It is important, therefore, that we begin our discussion of QoS with LANs.

Because of the migration to switched architectures in the LAN environment, the IEEE released a standard in late 1998 for the creation of *virtual LANs* (VLANs). A VLAN is nothing more than a single LAN that has been logically segmented into user groups or functional organizations. The elegance of the VLAN concept is that the members of a particular VLAN do not have to be physically collocated on the same network segment. They can be scattered all over the world, as long as they are part of the same corporate network.

QoS is necessary in the LAN environment to overcome the disparity that often exists between the bandwidth offered by the LAN and the capacity of the *wide area network* (WAN) with which it communicates. LAN bandwidth has always been greater than that of the WAN, and with the arrival of Fast and Gigabit Ethernet, the discrepancy between the two is widening at a rapid rate. Furthermore, the volume of LAN traffic is growing quickly, which means that the number of LANs contending for scarce WAN bandwidth is growing. Additionally, growth in IP-based networks, especially VPNs, is forcing that demand upward.

Several standards address the concerns of LAN QoS: 802.1Q and 802.1p. Both are discussed in the following sections.

802.1Q and 802.1p   The 802.1Q standard defines the interoperability requirements for vendors of LAN equipment wanting to offer VLAN capabilities. The standard was crafted to simplify the automation, configuration, and management of VLANs, regardless of the switch or end-station vendor.

Like MPLS, 802.1Q relies on the use of priority tags that indicate service classes within the LAN. These tags form part of the frame header and use three bits to uniquely identify eight service classes. These classes, as proposed by the IEEE, are shown in Table 1-1.

Closely associated with 802.1Q is 802.1p,[2] which enables the three QoS bits in the 802.1Q VLAN header to specify QoS requirements. It is primarily used by layer 2 bridges to filter and prioritize multicast

[2]The fact that the *Q* is uppercase and *p* is lowercase is not an accident. IEEE 802.1p is an adjunct standard and does not stand on its own. IEEE 802.1Q, on the other hand, is an independent standard.

Table 1-1

IEEE service
classes

Priority	Binary Value	Traffic Type
7	111	Network control
6	110	Interactive voice
5	101	Interactive multimedia
4	100	Streaming multimedia
3	011	Excellent effort
2	010	Spare
1	001	Background

traffic. The 802.1p QoS bits can be set by intelligence in the client machine as dictated by network policy established by the network management organization. In a practical application, 802.1p can be converted to the *Differential Services Working Group* (DiffServ) for QoS integration across the wide area. After all, 802.1p is really a QoS specification for LAN environments, most typically Ethernet. Therefore, the DiffServ byte in the IP header can be encoded at the edge of the network by the ingress router based on information contained in the 802.1p field in the Ethernet frame header. At the egress router, the opposite occurs, guaranteeing end-to-end QoS across the wide area.

Of course, these standards only address the requirements of LANs, which will inevitably interconnect with WAN protocols such as ATM or IP, which are discussed later in this section. Consequently, the IETF DiffServ committee has developed standards for interoperability between 802.1Q and wide area protocols such as IP's DiffServ, while the ATM Forum has a similar effort underway to map 802.1Q priority levels to ATM service classes.

DiffServ and MPLS The IETF has taken an active role in the development of QoS standards for IP-based transmission. Under their auspices, two working groups have emerged with responsibilities for

QoS issues. The first is DiffServ and the second is the *Multiprotocol Label Switching Working Group* (MPLS).

When establishing connections for *voice over IP* (VoIP), it is critical to manage queues to ensure the proper treatment of packets that derive from delay-sensitive services. In order to do this, packets must be differentiable—that is, voice and video packets must be identifiable so that they can be treated properly. Routers, in turn, need to be able to respond properly to delay-sensitive traffic by implementing queue management processes. This requires the routers to establish both normal and expedited queues, and to handle traffic in expedited routing queues faster than the arrival rate of the traffic. This translates into a traffic-policing requirement to ensure that the offered load remains below the bandwidth reserved at each node for high-priority data.

Differentiated Services (DiffServ) DiffServ and MPLS have the same goal in mind, but approach it from different directions. DiffServ has the capability to prioritize packets through the use of bits in the IP header known as the *Differential Services Code Point* (DSCP), formerly part of the *type of service* (TOS) field. It relies on *per-hop behaviors* (PHBs), which define the traffic characteristics that must be accommodated. The best known of these is the Expedited Forwarding PHB, designed to be used for services that require minimum delay and jitter such as voice and video. DiffServ, therefore, is a technique for classifying packets according to QoS requirements. Because the classification process occurs at the edge of the network, it scales well as the network grows.

DiffServ breaks the responsibilities of traffic management into four key areas based on the overall architecture of the network, as illustrated in Figure 1-41. At the customer's access router, traffic is managed according to flow requirements and clustered for delivery to the service provider's network. The traffic is then handed to the service provider's ingress router, which sits at the edge of the network and is responsible for implementing the *service level agreement* (SLA) between the customer and the service provider.

Once the edge routers have classified the traffic, the core routers can handle it according to the DSCP markers they assign to each packet. Within the network, then, core transit routers simply route

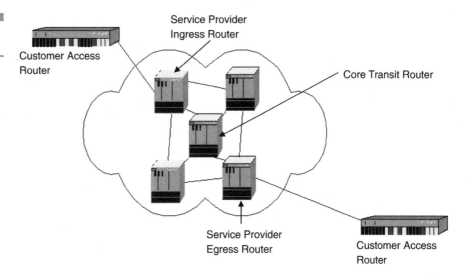

Service Provider
Ingress Router

Customer Access
Router

Core Transit Router

Service Provider
Egress Router

Customer Access
Router

the traffic as required. By the time the traffic arrives at the core, the edge devices have already classified it and the core router simply handles the interior routing function. Ultimately, the traffic reaches the service provider's egress router, which is another edge device that performs additional traffic-shaping functions to ensure compliance with the SLA. Thus, the core can be extremely fast because the classification process has already been done at the point of ingress.

DiffServ, therefore, is *not* an end-to-end protocol. Traffic, perhaps from a LAN, arrives at the edge of the network, where DiffServ's domain begins. The ingress and egress routers manage and shape traffic flows with the freedom to use packet discard if appropriate.

Multiprotocol Label Switching (MPLS) MPLS is considered to be superior to DiffServ, although both techniques rely on edge routers to classify and label the packets at the point of ingress. MPLS achieves the same goal as DiffServ by establishing virtual circuits known as *label-switched paths* (LSPs), which are built around specific QoS requirements. Thus, a router can establish LSPs with explicit QoS capabilities and route packets to those LSPs as required, guaranteeing the delay that a particular flow will encounter end to end. Some industry analysts have compared MPLS LSPs to the trunks that are established in the voice environment.

Although both MPLS and DiffServ offer reasonable levels of QoS control in IP environments, neither has the level of capability that ATM offers, which explains why ATM combined with IP is the choice of technologies for many players looking to deploy mixed services over IP. Both techniques offer services to the user, but they are lacking in a number of key QoS areas such as routing capability. Both DiffServ and MPLS describe techniques for classifying and labeling a variety of QoS levels, but neither of them speak to the requirements for establishing an end-to-end path that offers constant quality. So although both (and especially MPLS for its simplicity) are often compared to ATM, ATM is clearly a more robust and capable protocol. *Multiprotocol over ATM* (MPOA) is one technique that ATM uses to manage ingress traffic and identify flows that require diverse QoS levels. It establishes virtual channels for specific QoS between various devices within the network, and routes traffic as befits its unique requirements for service. If it finds traffic that requires specific handling that does not already have a virtual channel established, it creates a default channel to ensure delivery. Thus, ATM handles both packet classification and routing.

MPLS is similar to MPOA and uses a two-part process for routing. First, it divides the packets into various *Forwarding Equivalence Classes* (FECs) and then maps the FECs to their next hop point. This process is performed at the point of ingress. Each FEC is given a fixed-length label that accompanies each packet from hop to hop. At each router, the FEC label is examined and used to route the packet to the next hop point, where it is assigned a new label.

For videoconferencing applications, QoS becomes a major issue. If the conferencing is being conducted across a public IP network, the number of router-to-router hops that the packets must make as they travel from the source to the destination can have a devastating impact on the quality of the received signal. For this reason, videoconferencing technology solutions have emerged that strive to reduce the number of hops by caching content locally and using shortest-hop protocols such as *Open Shortest Path First* (OSPF).

Because of the demand for low-latency performance that video places on a network, many corporations do not mix the traffic on the LAN. Instead, they use a dual-port router that has the capability to segregate the video traffic from the data traffic that is less delay sensitive. So although both traffic types may be transported using IP, they do not mingle at the local network level to preserve the integrity of both. The bursty nature of the LAN could adversely affect the quality of the video signal, and the video, which is a bandwidth hog, could seriously affect the LAN. So most corporations keep them segregated.

We now turn our attention to access technologies, the network components that provide the interconnection between premise devices and the transport network.

For the longest time, "access" described the manner in which customers reached the network for the transport of voice services. In the last 20 years, however, that definition has changed dramatically. In 1981, IBM changed the world when it introduced the PC, and in 1984, the Macintosh arrived, bringing well-designed and organized computing power to the proverbial masses. Shortly thereafter, hobbyists began to take advantage of emergent modem technology and created online databases—the first bulletin board systems that allowed people to send simple text messages to each other. This accelerated the modem market dramatically, and before long, data became a common component of local loop traffic. At that time, there was no concept of Instant Messenger or of the degree to which e-mail would fundamentally change the way people communicate and do business. At the same time, the business world found more and more applications for data, and the need to move that data from place to place became a major contributor to the growth in data traffic on the world's telephone networks. Today, video, videoconferencing, and yes, even voice—mostly from audio bridges used for conferencing—are major contributors to the overall traffic volume as well.

In those heady, early days, data did not represent a problem for the bandwidth-limited local loop. The digital information created by a computer and intended for transmission through the telephone network was received by a modem, converted into a modulated analog waveform that fell within the 4 KHz voiceband, and fed to the network without incident.

Over time, modem technology advanced, enabling the local loop to provide higher bandwidth. This increasing bandwidth was made possible through clever signaling schemes that enable a single signaling event to transport more than a single bit. These modern modems are commonplace today. They allow baud levels to reach unheard-of extremes and permit the creation of very high bit-per-signal rates.

The analog local loop is used today for various voice and data applications in both business and residence markets. The new lease on life it enjoys thanks to advanced modem technology as well as a focus by installation personnel on the need to build a clean, reliable outside plant has resulted in the development of faster access technologies designed to operate across the analog local loop, including traditional high-speed modem access and options such as DSL.

Marketplace Realities

According to a number of demographic studies conducted in the last 18 months, there are approximately 70 million households today that host home office workers, and the number is growing rapidly. These numbers include both telecommuters and those who are self-employed and work out of their homes. They require the ability to connect to remote LANs and corporate databases, retrieve e-mail, access the Web, and conduct videoconferences with colleagues and customers. The traditional bandwidth-limited local loop is not capable of satisfying these requirements with traditional modem technology. Dedicated private-line service, which would solve the problem, is far too expensive as an option, and because it is dedicated, it is not particularly efficient. Other solutions are required, and these have emerged in the form of access technologies that take advantage of either a conversion to end-to-end digital connectivity (ISDN) or expanded capabilities of the traditional analog local loop (DSL and 56K modems). In some cases, a whole new architectural approach is causing excitement in the industry (*Wireless Local Loop* [WLL]). Finally, cable access has become a popular option as the cable infrastructure has evolved to a largely optical, all-digital system with high-bandwidth, two-way capabilities. We will discuss each of these options in the pages that follow.

56 Kbps Modems

One of the most important words in telecommunications is "virtual." It is used in a variety of ways, but in reality, it only has one meaning. If you see the word "virtual" associated with a technology or product, you should immediately say to yourself, "It's a lie."

A 56 Kbps modem is a good example of a virtual technology. These devices have attracted a great deal of interest since they were introduced a few years ago. Under certain circumstances, they do offer higher access speeds designed to satisfy the increasing demands of bandwidth-hungry applications and increasingly graphics-oriented web pages. The problem they present is that they do not really provide true 56K access, even under the best of circumstances.

56K modems provide asymmetric bandwidth, with 56 Kbps delivered downstream toward the customer (sometimes) and significantly less bandwidth (33.6 Kbps) in the upstream direction. Although this may seem odd, it makes sense given the requirements of most applications today that require modem access. A web session, for example, requires very little bandwidth in the upstream direction to request that a page be downloaded. The page itself, however, may require significantly more because it may be replete with text, graphics, Java applets, and even small video clips. Because the majority of modem access today is for Internet surfing, asymmetric access is adequate for most users.

The limitations of 56K modems stem from a number of factors. One of them is the fact that under current FCC regulations (specifically Part 68), line voltage supplied to a communications facility is limited such that the maximum achievable bandwidth is 53 Kbps in the downstream direction. Another limitation is that these devices require that only a single analog-to-digital conversion occur between the two endpoints of the circuit. This typically occurs on the downstream side of the circuit and usually at the interface point where the local loop leaves the central office. Consequently, downstream traffic is less susceptible to the noise created during the analog-to-digital conversion process, while the upstream channel is affected by it and is therefore limited in terms of the maximum bandwidth it can provide. In effect, 56K modems, in order to achieve their maximum bandwidth, require that one end of the circuit, typically the central office end, be digital.

The good news with regard to 56K modems is that even in situations where the 56 Kbps speed is not achievable, the modem will fall back to whatever maximum speed it can fulfill. Furthermore, no premises wiring changes are required, and because this device is really nothing more than a faster modem, the average customer is comfortable with migration to the new technology. This is certainly demonstrated by sales volume. As mentioned before, most PCs today are automatically shipped with a 56K modem.

V.92

Even at 56 Kbps, V.90 modems had to put up with users grousing about the asymmetric nature of data transport supported by the modem (53 Kbps downstream and 33.6 Kbps upstream) and the fact that while connected, incoming calls were inaccessible. First announced in July 2000 and now becoming available, a new set of modems based on the V.92 standard is allaying those concerns. V.92 adds three new capabilities that expand upon the capabilities offered by V.90 modems: Quick Connect, Modem-on-Hold, and PCM Upstream.

Many in the industry have argued that the growing deployment of ISDN, DSL, and cable modems would obviate the need for another analog modem standard, even with its enhanced capabilities. As we will see a bit later, none of these relatively new technologies have enjoyed raging successes, the result of which is that most industry authorities believe that as many as 55 percent of Internet users will still be using analog modem access technologies in 2004. The need for V.92 is therefore justified.

The three added capabilities that V.92 brings to the modem table are significant. The first of these, Quick Connect, does precisely what the name implies—it reduces the typical connect time by 50 percent or more, shrinking the average 25-second connection to as little as 10 seconds. It does this in a rather clever fashion. Part of the modem handshake process involves an assessment of the characteristics of the transmission facility to determine the maximum speed at which the two modems can communicate. Most users make multiple calls to the same number over the same facility, even while traveling (from a hotel room, for example). The characteristics of the line rarely change between calls so V.92 modems take advantage of this fact by memorizing the line characteristics to reduce the length of successive handshake procedures.

The second new service is called Modem-on-Hold. This feature enables a user to receive an incoming call while maintaining a previously established connection to the Internet. The service requires that the standard telco-provided call-waiting feature be available on the line, but this is a trivial issue. The service also works in reverse: a user can maintain a modem connection while initiating a voice call.

The third feature, PCM Upstream, addresses the complaint of asymmetric transport. PCM Upstream increases the upstream data rate from 33.6 Kbps to a maximum of 48 Kbps, a 30 percent increase. It has no effect on the download speed, but most user complaints have been with regard to upstream connectivity. Furthermore, a new compression standard, V.44, improves the service provided by traditional V.42bis compression by as much as 200 percent for compressible data streams. This results in an overall throughput of as high as 300 Kbps.

All of these features taken together result in a significant service improvement over V.90 modems. Given the deployment and service issues that cable, ISDN, and DSL have suffered through, analog modems may be around longer than we currently believe. They do, after all, work and can be (within reason) universally deployed.

Of course, those of us who have tried to use dial-up technologies for videoconferencing know how well (or not) it works for such a bandwidth-intensive and quality-sensitive service. Other solutions are needed. We begin with ISDN.

ISDN

ISDN has been the proverbial technological roller coaster since its arrival as a concept in the late 1960s. Often described as "the technology that took 15 years to become an overnight success," ISDN's level of success has been all over the map. Internationally, it has enjoyed significant uptake as a true, digital local loop technology. In the United States, however, because of competing and often incompatible hardware implementations, high cost, and spotty availability, its deployment has been erratic at best. In market areas where providers have made it available at reasonable prices, it has been quite successful. Furthermore, it is experiencing something of a renaissance today because of DSL's failure to capture the market. ISDN is tested, well-understood, available, and fully functional. It currently offers 128 Kbps of bandwidth, while DSL's capability to do so is less certain.

Interestingly enough, in spite of the perception of failure that ISDN has been tarnished with over the years, it has always been a mainstay in videoconferencing.

ISDN Technology The typical non-ISDN local loop is analog. Voice traffic is carried from an analog telephone to the central office using a frequency-modulated carrier; once at the central office, the signal is typically digitized for transport within the digital network cloud. On one hand, this is good because it means that there is a digital component to the overall transmission path. On the other hand, the loop is still analog, and as a result, the true promise of an end-to-end digital circuit cannot be realized. The circuit is only as good as the weakest link in the chain, and the weakest link is clearly the analog local loop.

In ISDN implementations, local switch interfaces must be modified to support a digital local loop. Instead of using analog frequency modulation to represent voice or data traffic carried over the local loop, ISDN digitizes the traffic at the origination point, either in the voice set itself or in an adjunct device known as a *terminal adapter* (TA). The digital local loop then uses *time-division multiplexing* (TDM) to create multiple channels over which the digital information is transported that provide for a wide variety of truly integrated services.

The Basic Rate Interface (BRI) There are two well-known implementations of ISDN. The most common (and the one intended primarily for residence and small business applications) is called the

Basic Rate Interface (BRI). In BRI, the 2-wire local loop supports a pair of 64 Kbps digital channels known as B-Channels as well as a 16 Kbps D-Channel, which is primarily used for signaling but can also be used by the customer for low-speed (up to 9.6 Kbps) packet data. The B-Channels can be used for voice and data, and can be bonded together in some implementations to create a single 128 Kbps channel for videoconferencing or other higher-bandwidth applications.

Figure 1-42 shows the layout of a typical ISDN BRI implementation, including the alphabetic reference points that identify the regions of the circuit and the generic devices that make up the BRI. In this diagram, the LE is the local exchange, or switch. The NT-1 is the network termination device that serves as the demarcation point between the customer and the service provider; among other things, it converts the two-wire local loop to a four-wire interface on the customer's premises. The TE1 (terminal equipment, type 1) is an ISDN-capable device such as an ISDN telephone. This simply means that the phone is a digital device and is therefore capable of performing the voice digitization itself. A TE2 (terminal equipment, type 2) is a non-ISDN-capable device, such as a *Plain Old Telephone Service* (POTS) telephone. In the event that a TE2 is used, a TA must be inserted between the TE2 and the NT-1 to perform analog-to-digital conversion and rate adaptation.

The reference points mentioned earlier identify circuit components between the functional devices just described. The U reference point is the local loop, the S/T reference point sits between the NT-1 and the TEs, and the R reference point is found between the TA and the TE2.

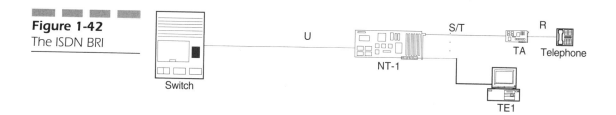

Figure 1-42

The ISDN BRI

BRI Applications Although BRI does not offer the stunning bandwidth that other more recent technologies (such as DSL) do, its bondable 64 Kbps channels provide reasonable capacity for many applications. The most common of these applications today are videoconferencing, remote LAN access, and Internet access. For the typical remote worker, the bandwidth available through BRI is more than adequate, and new video compression technology puts reasonable quality videoconferencing within the grasp of the end user at an affordable price. The 64 Kbps channels make short shrift of LAN-based text file downloads and reduce the time required for graphics-intensive web page downloads to reasonable levels.

The Primary Rate Interface (PRI) The other major implementation of ISDN is called the *Primary Rate Interface* (PRI). The PRI is really nothing more than a T-carrier in that it is a 4-wire local loop, uses AMI and B8ZS for ones-density control and signaling, and provides 24 to 64 Kbps channels that can be distributed among a collection of users as the customer sees fit (see Figure 1-43). In PRI, the signaling channel operates at 64 Kbps (unlike the 16 Kbps D-Channel in the BRI) and is not accessible by the user. It is used solely for signaling purposes—in other words, it cannot carry user data. The primary reason for this is service protection. In the PRI, the D-Channel is used to control the 23 B-Channels and therefore requires significantly more bandwidth than the BRI D-Channel. Furthermore, the PRI standards enable multiple PRIs to share a single D-Channel, which makes the D-Channel's operational consistency even more critical.

The functional devices and reference points are not appreciably different from those of the BRI. The local loop is still identified as the U reference point. In addition to an NT-1, we now add an NT-2, which is a service distribution device, usually a *private branch*

Figure 1-43
The ISDN PRI

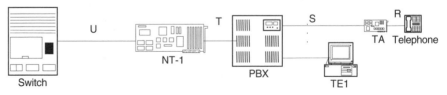

exchange (PBX), which allocates the PRI's 24 channels to customers. This makes sense because PRIs are typically installed at businesses that employ PBXs for voice distribution. The S/T reference point is now divided—the S reference point sits between the NT-2 and TEs, while the T reference point is found between the NT-1 and the NT-2.

 PRI service also has the capability to provision B-Channels as superrate channels to satisfy the bandwidth requirements of higher-bit-rate services. These are called H-Channels and are provisioned as shown in Table 1-2.

PBX Applications

The PRI's marketplace is the business community and its primary advantage is pair gain—that is, to conserve copper pairs by multiplexing the traffic from multiple user channels onto a shared, four-wire circuit. Because a PRI can deliver the equivalent of 23 voice channels to a location over a single circuit, it is an ideal technology for a number of applications, including the interconnection of a PBX to a local switch, a dynamic bandwidth allocation for higher-end videoconferencing applications, and the interconnection between an ISP's network and that of the local telephone company.

Some PBXs are ISDN-capable on the line (customer) side, meaning that they have the capability to deliver ISDN services to users who emulate the services that would be provided over a direct connection to an ISDN-provisioned local loop. On the trunk (switch) side, the PBX is connected to the local switch via one or more T1s,

Table 1-2

ISDN Superrate Services (H-channels)

Channel	Bandwidth
H0	384 Kbps (6B)
H10	1.472 Mbps (23B)
H11	1.536 Mbps (24B)
H12	1.920 Mbps (30B)

which in turn provide access to the telephone network. This arrangement results in significant savings, faster call setup, more flexible administration of trunk resources, and the ability to offer diverse services through the granular allocation of bandwidth as required. Recently, a few PBX manufacturers have made noises about PBXs that are fully equipped for IP conferencing and VoIP and are capable of assigning bandwidth on demand as required by the application. This market will grow in importance; watch it closely.

Digital Subscriber Line (DSL)

The access technology that has enjoyed the greatest amount of attention in recent times is DSL. It provides a good solution for remote LAN access, Internet surfing, and access for telecommuters to corporate databases.

DSL came about largely as a direct result of the Internet's success. Prior to its arrival, the average telephone call lasted approximately four minutes, a number that central office personnel used while engineering switching systems to handle expected call volumes. This number was arrived at after nearly 125 years of experience designing networks for voice customers. They knew about Erlang theory, loading objectives, peak-calling days/weeks/seasons, and had decades of trended data to help them anticipate load problems. So in 1993, when the Internet—and with it, the World Wide Web—arrived, network performance became unpredictable as callers began to surf, often for hours on end. The average four-minute call became a thing of the past as online service providers such as *America Online* (AOL) began to offer flat-rate plans that did not penalize customers for long connect times. Then, the unthinkable happened: switches in major metropolitan areas, faced with unpredictably high call volumes and hold times, began to block during normal business hours, a phenomenon that only occurred in the past during disasters or on Mother's Day.

A number of solutions were considered, including charging different rates for data calls than for voice calls, but none of these proved feasible until a technological solution was proposed. The solution was DSL.

It is commonly believed that the local loop is incapable of carrying more than the frequencies required to support the voiceband. This is a misconception. ISDN, for example, requires significantly high bandwidth to support its digital traffic. When ISDN is deployed, the loop must be modified in certain ways to eliminate its designed-in bandwidth limitations. For example, some long loops are deployed with load coils that "tune" the loop to the voiceband. They make the transmission of frequencies above the voiceband impossible, but allow the relatively low-frequency voiceband components to be carried across a long local loop. High-frequency signals tend to deteriorate faster than low-frequency signal components, so the elimination of the high frequencies extends the transmission distance and reduces transmission impairments that would result from the uneven deterioration of a rich, multifrequency signal. These load coils, therefore, must be removed if digital services are to be deployed.

A local loop is only incapable of transporting high-frequency signal components if it is designed not to carry them. The capacity is still there; the network design, through load coil deployment, makes that additional bandwidth unavailable. DSL services, especially *Asymmetric Digital Subscriber Line* (ADSL), take advantage of this disguised bandwidth.

DSL Technology In spite of the name, Digital Subscriber Line (DSL) is an analog technology. The devices installed on each end of the circuit are sophisticated high-speed modems that rely on complex encoding schemes to achieve the high bit rates that DSL offers. Furthermore, several of the DSL services, specifically ADSL, g.lite, *Very High-Speed Digital Subscriber Line* (VDSL), and *Rate-Adaptive Digital Subscriber Line* (RADSL) are designed to operate in conjunction with voice across the same local loop. ADSL is the most commonly deployed service and offers a great deal to both business and residence subscribers.

XDSL Services DSL comes in a variety of flavors designed to provide flexible, efficient, high-speed service across the existing telephony infrastructure. From a consumer's point of view, DSL, especially ADSL, offers a remarkable leap forward in terms of available bandwidth for broadband access to the Web. As content has steadily moved away from being largely text based and has become more graphical, the demand for faster delivery services has grown for some time now. DSL may provide the solution at a reasonable cost to both the service provider and the consumer.

Businesses will also benefit from DSL. Remote workers, for example, can rely on DSL for LAN and Internet access. Furthermore, DSL provides a good solution for VPN access as well as for ISPs looking to grow the bandwidth available to their customers. It is available in a variety of both symmetric and asymmetric services and therefore offers a high-bandwidth access solution for a variety of applications. The most common DSL services are ADSL, *High-Bit-Rate Digital Subscriber Line* (HDSL), HDSL-2, RADSL, and VDSL. The special case of g.lite, a form of ADSL, will also be discussed.

Asymmetric Digital Subscriber Line (ADSL)   When the World Wide Web and flat-rate access charges arrived, the typical consumer phone call went from roughly four minutes in duration to several times that. All the engineering that led to the overall design of the network based on an average four-minute hold time went out the window as the switches staggered under the added load. Never was the expression "In its success lie the seeds of its own destruction" more true. When ADSL arrived, it provided the offload required to save the network.

A typical ADSL installation is shown in Figure 1-44. No change is required to the two-wire local loop; minor equipment changes, however, are required. First, the customer must have an ADSL modem at his or her premises. This device enables the telephone service and a data access device, such as a PC, to be connected to the line.

The ADSL modem is more than a simple modem in that it also provides the *frequency division multiplexing* (FDM) process required to separate the voice and data traffic for transport across the loop. The device that actually does this, as shown in Figure 1-45, is called a *splitter* because it splits the voice traffic away from the data. It is usually bundled as part of the ADSL modem, although it can also be installed as a card in the PC, as a stand-alone device at the demarcation point, or on each phone at the premises. The most common implementation is to integrate the splitter as part of the DSL modem; this, however, is the least desirable implementation because

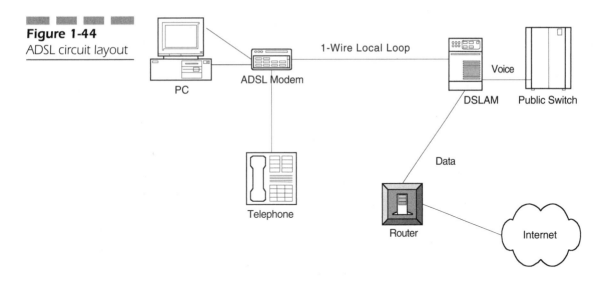

Figure 1-44
ADSL circuit layout

Figure 1-45
An ADSL splitter

this design can lead to crosstalk between the voice and data circuitry inside the device. When voice traffic reaches the ADSL modem, it is immediately encoded in the traditional voiceband and handed off to the local switch when it arrives at the central office. The modem is often referred to as an *ADSL Transmission Unit for Remote* (ATU-R) use. Similarly, the device in the central office is often called an ATU-C (for central office use).

When a PC wants to transmit data across the local loop, the traffic is encoded in the higher-frequency band reserved for data traffic. The ADSL modem knows to do this because the traffic is arriving on a port reserved for data devices. Upon arrival at the central office, the data traffic does not travel to the local switch; instead, it stops at the ADSL modem that has been installed at the central office end of the circuit. In this case, the device is actually a bank of modems that serves a large number of subscribers and is known as a *Digital Subscriber Line Access Multiplexer* (DSLAM) (pronounced "dee-slam").

Instead of traveling onto the local switch, the data traffic is now passed around the switch to a router, which in turn is connected to the Internet. This process is known as a *line-side redirect*.

The advantages of this architecture are fairly obvious. First, the redirect offloads the data traffic from the local switch so that it can go back to doing what it does best—switching voice traffic. Second, it creates a new line of business for the service provider. As a result of adding the router and connecting the router to the Internet, the service provider instantly becomes an ISP. This is a near-ideal combination because it enables the service provider to become a true service provider by offering much more than simple access and transport.

As the name implies, ADSL provides two-wire asymmetric service —that is, the upstream bandwidth is different from the downstream. In the upstream direction, data rates vary from 16 to 640 Kbps, while the downstream bandwidth varies from 1.5 to 8 Mbps. Because most applications today are asymmetric in nature, this disparity poses no problem for the average consumer of the service. Of course, video is not. It requires symmetric service and must therefore see equivalent levels of high-bandwidth transport in both upstream and downstream directions. ADSL, unless it operates consistently at the highest level of its operating range, is probably not adequate for most videoconferencing applications.

A Word About the DSLAM This device has received a significant amount of attention recently because of the central role that it plays in the deployment of broadband access services. Obviously, the

DSLAM must interface with the local switch so that it can pass voice calls onto the PSTN. However, it often interfaces with a number of other devices as well. For example, on the customer side, the DSLAM can connect to a standard ATU-C directly to a PC with a built-in NIC, to a variety of DSL services, or to an integrated access device of some kind. On the trunk side (facing the switch), the DSLAM can connect to IP routers as described before, to an ATM switch, or to some other broadband service provider. Therefore, it becomes the focal point for the provisioning of a wide variety of access methods and service types.

High-Bit-Rate Digital Subscriber Line (HDSL)   The greatest promise of HDSL is that it provides a mechanism for the deployment of four-wire T1 and E1 circuits without the need for span repeaters, which can significantly add to the cost of deploying data services. It also means that service can be deployed in a matter of days rather than weeks, something customers certainly applaud.

DSL technologies in general allow for repeaterless facilities as far as 12,000 feet, whereas traditional four-wire data circuits such as T1 and E1 require repeaters every 6,000 feet. Consequently, many telephone companies are now using HDSL behind the scenes as a way to deploy these traditional services. Customers do not realize that the T1 facility they are plugging their equipment into is being delivered using HDSL technology. The important thing is that they don't need to know. All the customer should have to care about is that there is now a SmartJack installed in the basement and through that jack they have access to 1.544 or 2.048 Mbps of bandwidth—period.

HDSL-2   HDSL-2 offers the same service that HDSL offers with one added (and significant) advantage. It does so over a single pair of wire, rather than two. It also provides other advantages. First, it was designed to improve vendor interoperability by requiring less equipment at either end of the span (transceivers or repeaters). Second, it was designed to work within the confines of a standard telephone company's *Carrier Serving Area* (CSA) guidelines by offering a 12,000-foot wire-run capability that matches the requirements of CSA deployment strategies.

A number of companies have deployed T1 access over HDSL-2 at rates 40 percent lower than typical T-carrier prices. Furthermore, a number of vendors including 3Com, Lucent, Nortel Networks, and Alcatel have announced their intent to work together to achieve interoperability among DSL modems.

Both HDSL and HDSL-2 are perfect solutions for videoconferencing because of the symmetric nature of their bandwidth-rich channels. In areas where they are available, they have enjoyed significant levels of deployment.

Rate-Adaptive Digital Subscriber Line (RADSL) RADSL (pronounced "rad-zel") is a variation of ADSL designed to accommodate changing line conditions that can affect the overall performance of the circuit. Like ADSL, it relies on DMT encoding, which selectively populates subcarriers with transported data, thus allowing for granular rate setting.

Very High-Speed Digital Subscriber Line (VDSL) VDSL is the newest DSL entrant in the bandwidth game and shows promise as a provider of extremely high levels of access bandwidth—as much as 52 Mbps over a short local loop. VDSL requires *Fiber-to-the-Curb* (FTTC) architecture and recommends ATM as a switching protocol. From a fiber hub, copper tail circuits deliver the signal to the business or residential premises. Bandwidth available through VDSL ranges from 1.5 to 6 Mbps on the upstream side and from 13 to 52 Mbps on the downstream side. Obviously, the service is distance sensitive and the actual achievable bandwidth drops as a function of distance. Nevertheless, even a short loop is respectable when such high bandwidth levels can be achieved. With VDSL, 52 Mbps can be reached over a loop length of up to 1,000 feet, which is not an unreasonable distance by any means.

G.lite Because the installation of splitters proved to be a contentious and problematic issue, the need arose for a version of ADSL that did not require them. That version is known as either *ADSL Lite* or *G.lite* (after the ITU-T G-Series standards that govern much of the ADSL technology). In 1997, Microsoft, Compaq, and Intel created the *Universal ADSL Working Group* (UAWG),[3] an organization that grew to nearly 50 members dedicated to the task of simplifying the rollout of ADSL. In effect, the organization had four stated goals:

- To ensure that analog telephone service will work over the g.lite deployment without remote splitters, in spite of the fact that the quality of the voice may suffer slightly due to the potential for impedance mismatches.

- To maximize the length of deployed local loops by limiting the maximum bandwidth provided. Research indicates that customers are far more likely to notice a performance improvement when migrating from 64 Kbps to 1.5 Mbps than when going from 1.5 Mbps to higher speeds. Perception is clearly important in the marketplace so the UWAG chose 1.5 Mbps as their downstream speed.

- To simplify the installation and use of ADSL technology by making the process as plug-and-play as possible.

- To reduce the cost of the service to a perceived reasonable level.

Of course, G.lite is not without its detractors. A number of vendors have pointed out that if G.lite requires the installation of microfilters at the premises on a regular basis, then true splitterless DSL is a myth because microfilters are in effect a form of splitter. They contend that if the filters are required anyway, then they might as well be used in full-service ADSL deployments to guarantee high-quality service delivery. Unfortunately, this flies in the face of one of the key tenets of G.lite, which is to simplify and reduce the cost of DSL deployment by eliminating the need for an installation dispatch (a "truck roll" in the industry's parlance). The key to G.lite's success in the eyes of the implementers is to eliminate the dispatch, minimize the impact on traditional POTS telephones, reduce costs, and extend the achievable drop length. Unfortunately, customers still have to be burdened with the installation of microfilters, and coupled noise on

---

[3]The group self-dissolved in the summer of 1999 after completing what they believed their charter to be.

POTS is higher than expected. Many vendors argue that these problems largely disappear with full-feature ADSL using splitters. A truck dispatch is still required, but again, it is often required to install the microfilters anyway so there is no net loss. Furthermore, a number of major semiconductor manufacturers support both G.lite and ADSL on the same chipset so the decision to migrate from one to the other is a simple one that does not necessarily involve a major replacement of internal electronics.

DSL Market Issues   DSL technology offers advantages to the service provider and the customer. The service provider benefits from successful DSL deployment because it serves not only as a cost-effective technique for satisfying the bandwidth demands of customers in a timely fashion, but it also provides a Trojan horse approach to the delivery of certain preexisting services. As we noted earlier, many providers today implement T1 and E1 services over HDSL because it offers a cost-effective way to do so. Customers are blissfully unaware of the fact; in this case, it is the service provider rather than the customer who benefits most from the deployment of the technology. From a customer's point of view, DSL provides a cost-effective way to buy medium to high levels of bandwidth, and in some cases, embedded access to content. Videoconferencing is one of many applications that stand to gain from the deployment of these technologies.

The Cable Network

The traditional cable network is an analog system based on a tree-like architecture. The headend, which serves as the signal origination point, serves as the signal aggregation facility. It collects programming information from a variety of sources including satellite and terrestrial feeds. Headend facilities often look like a mushroom farm; they are typically surrounded by a variety of satellite dishes (see Figures 1-46 and 1-47).

Figure 1-46
Satellite dishes

Figure 1-47
Smaller satellite
dishes

The headend is connected to the downstream distribution network by a one-inch-diameter rigid coaxial cable, as shown in Figure 1-48. That cable delivers the signal, usually a 450 MHz collection of 6 MHz channels, to a neighborhood, where splitters divide the signal and send it down a half-inch diameter semirigid coax that typically runs down a residential street. At each house, another splitter (see Figure 1-49) pulls off the signal and feeds it to the set-top box in the house over the drop wire, a local loop of flexible quarter-inch coaxial cable.

Although this architecture is perfectly adequate for the delivery of one-way television signals, its shortcomings for other services should be fairly obvious to the reader. First of all, it is, by design, a broadcast system. It does not typically have the capability to support upstream traffic (from the customer toward the headend) and is therefore not suited for interactive applications like videoconferencing. Second, because of its design, the network is prone to significant failures that have the potential to affect large numbers of customers. The tree structure, for example, means that if a failure occurs along any branch in the tree, every customer from that point downward loses service. Contrast this with the telephone network where customers have a dedicated local loop over which their service is delivered. Second, because the system is analog, it relies on amplifiers to keep the

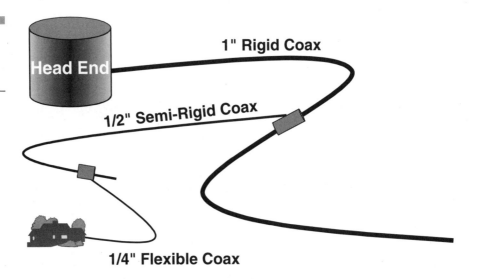

Figure 1-48
Layout of a typical cable distribution network

Figure 1-49

Signal splitter in
residential cable
installation

signal strong as it is propagated downstream. These amplifiers are powered locally—they do not have access to central office power as devices in the telephone network do. Consequently, a local power failure can bring down the network's capability to distribute service in that area.

The third issue is one of customer perception. For any number of reasons, there is a general perception that the cable network is not as capable or as reliable as the telephone network. As a consequence of this perception, the cable industry is faced with the daunting challenge of convincing potential voice and data customers that they are in fact capable of delivering high-quality service. Some of the concerns are justified. In the first place, the telephone network has been in existence for almost 125 years, during which time its operators have learned how to optimally design, manage, and operate it in order to provide the best possible service. The cable industry, on the other hand, came about 50 years ago and didn't benefit from the rigorously administered, centralized management philosophy that characterized the telephone industry. Additionally, the typical 450 MHz cable system did not have adequate bandwidth to support the bidirectional transport requirements of new services.

Furthermore, the architecture of the legacy cable network, with its distributed power delivery and tree-like distribution design, does not lend itself to the same high degree of redundancy and survivability that the telephone network offers. Consequently, cable providers have been hard-pressed to convert customers who are vigorously protective of their telecommunications services.

Evolving Cable Systems   Faced with these harsh realities and the realization that the existing cable plant could not compete with the telephone network in its original analog incarnation, cable engineers began a major rework of the network in the early 1990s. Beginning with the headend and working their way outward, they progressively redesigned the network to the extent that in many areas of the country their coaxial local loop is capable of competing on equal footing with the telco's twisted pair—and in some cases, beating it.

The process they have used in their evolution consists of four phases. In the first phase, they converted the headend from analog to digital. This enabled them to digitally compress the content, resulting in a far more efficient utilization of the available bandwidth. Second, they undertook an ambitious physical upgrade of the coaxial plant, replacing the one-inch trunk and half-inch distribution cable with optical fiber. This brought about several desirable results. First, by using a fiber feeder, network designers could eliminate a significant number of the amplifiers responsible for the failures the network experienced due to power problems in the field. Second, the fiber makes it possible to provision significantly more bandwidth than coaxial systems allow. Third, because the system is digital, it suffers less from noise-related errors than its analog predecessor did. Finally, an upstream return channel was provisioned, as shown in Figure 1-50, which makes the delivery of true interactive services such as voice, web surfing, and videoconferencing possible. Unfortunately, from the point of view of would-be videoconferencing users, cable is its own worst enemy. Because of its growing popularity, the number of cable modem users has grown rapidly. Because cable is a shared architecture, because there are many multimedia applications users, and because there is finite bandwidth in the system, the amount of bandwidth available to each user shrinks as the number of users climbs. This is particularly vexing for users whose applications require constant bandwidth if they are to perform properly—like videoconferencing. There is discussion now among cable engineers about how they can create symmetric transport systems for their data users, but there are currently no concrete plans. Stay tuned.

The third phase of the next-generation cable system conversion had to do with the equipment provisioned at the user's premises. The analog set-top box has now been replaced with a digital device that has the capability to take advantage of the capabilities of the network, including access to the upstream channel. It decompresses digital content, performs content (*stuff*) separation, and provides the network interface point for data and voice devices.

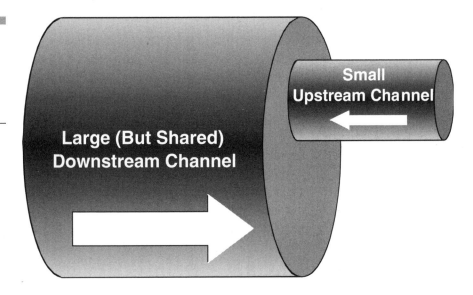

Figure 1-50
Upstream and downstream channels in modern cable systems

Small Upstream Channel

Large (But Shared) Downstream Channel

The final phase is business conversion. Cable providers look forward to the day when their networks will compete on equal footing with the twisted-pair networks of the telephone company, and customers will see them as viable competitors. In order for this to happen, they must demonstrate that their network is capable of delivering a broad variety of competitive services, that the network is robust, that they have *operations support systems* (OSSs) that will guarantee the robustness of the infrastructure, and that they are financially competitive with incumbent providers. They must also create a presence for themselves in the business centers of the world. Today they are almost exclusively residential service providers. If they are to break into the potentially lucrative business market, they must have a presence there.

Cable Modems   As cable providers have progressively upgraded their networks to include more fiber in the backbone, their plan to offer two-way access to the Internet has become a reality. Cable modems offer access speeds of up to 10 Mbps, and so far the market uptake has been spectacular.

Cable modems provide an affordable option to achieve high-speed access to the Web, with current monthly subscription rates in the neighborhood of $40. They offer asymmetric access, that is, a much higher downstream speed than upstream; however, for the majority of users, this does not represent a problem since the bulk of their use will be for web surfing during which the bulk of the traffic travels in the downstream direction anyway.

Although cable modems speed up access to the Web and other online services by several orders of magnitude, a number of downsides must be considered. The greatest concern that has been voiced about cable modems is security. Because cable modems are always on, they represent an easy entry point for hackers looking to break into machines. It is therefore critical that cable subscribers use some form of firewall software or a router that has the capability to perform filtering.

Data over Cable Standards   As interest grew in the late 1990s for broadband access to data services over cable television networks,

CableLabs®, working closely with the ITU and major hardware vendors, crafted a standard known as the *Data over Cable Service Interface Specification* (DOCSIS). The standard is designed to ensure interoperability among cable modems as well as to assuage concerns about data security over shared cable systems. DOCSIS has done a great deal to resolve marketplace issues.

Under the standards, CableLabs crafted a cable modem certification standard called DOCSIS 1.0 that guarantees that modems carrying the certification will interoperate with any headend equipment, are ready to be sold in the retail market, and will interoperate with other certified cable modems. Engineers from Askey, Broadcom, Cisco Systems, Ericsson, General Instrument, Motorola, Philips, 3Com, Panasonic, Digital Furnace, Thomson, Terayon, Toshiba, and Com21 participated in the development effort.

The DOCSIS 1.1 specification was released in April 1999 and included two additional functional descriptions that began to be implemented in 2000. The first specification details procedures for guaranteed bandwidth as well as a specification for QoS guarantees. The second specification is called *Baseline Privacy Interface Plus* (BPI+). It enhances the current security capability of the DOCSIS standards through the addition of digital-certificate-based authentication and support for multicast services to customers.

Although the DOCSIS name is in widespread use, CableLabs now refers to the overall effort as the *CableLabs Certified Cable Modem Project*.

Wireless Access Technologies

It is only in the last few years that wireless access technologies have advanced to the point where they are being taken seriously as contenders for the broadband local loop market. Traditionally, there was minimal infrastructure in place, and it was largely bandwidth-bound and error-prone to the point that wireless solutions were not considered serious contenders.

Wireless Access

To understand wireless communications, it is necessary to examine both radio and telephone technologies because the two are inextricably intertwined. In 1876, Alexander Graham Bell, a part-time inventor and a teacher of hearing-impaired students, invented the telephone while attempting to resolve the challenge of transmitting multiple telegraph signals over a shared pair of wires. His invention changed the world forever.

In 1896, a mere 20 years later, Italian engineer and inventor Guglielmo Marconi developed the spark gap radio transmitter, which eventually enabled him to transmit long-wave radio signals across the Atlantic Ocean as early as 1901. Like Bell, his invention changed the world. For his contributions, he was awarded the Nobel Prize in 1909.

It wasn't until the 1920s, though, when these two technologies began to dovetail, that their true promise was realized. Telephony provided interpersonal, two-way, high-quality voice communications, but required the user to be stationary. Radio, on the other hand, provided mobile communications, but was limited by distance, environmentally induced signal degradation, and spectrum availability. Whereas telephony was advertised as a universally available service, radio was more of a catch-as-catch-can offering that was subject to severe blocking. If a system could be developed that combined the signal quality and ubiquity of telephony with the mobility of radio, however, a truly promising new service offering could be made available.

Today, cellular telephony (and other services like it) provides high-quality, almost ubiquitous, wireless telephone service. Thanks to advances in digital technology, wireless telephony also offers services that are identical to those provided by the wired network. The pricing for wired and wireless services is now reaching parity. Flat-rate nationwide pricing models that have no roaming or long-distance restrictions are commonplace today.

With the arrival of *third-generation* (3G) wireless systems, there is now significant discussion afoot about the ability to deliver broadband multimedia transport to wireless devices such as cell phones or PDAs. In Japan, video to the phone is already appearing on a limited basis, but it is not yet considered to be a primetime application. The technology will undoubtedly come about, but not for some time.

Local Multipoint Distribution Service (LMDS)   LMDS is a bottleneck resolution technology that is designed to alleviate the transmission restriction that occurs between high-speed LANs and WANs. Today, local networks routinely operate at speeds of 100 Mbps (Fast Ethernet) and even 1,000 Mbps (Gigabit Ethernet), which means that any local loop solution that operates slower than either of those poses a restrictive barrier to the overall performance of the system. LMDS offers a good alternative to wired options. Originally offered as CellularVision, it was seen by its inventor, Bernard Bossard, as a way to provide cellular television as an alternative to cable.

Operating in the 28 GHz range, LMDS offers data rates as high as 155 Mbps, the equivalent of SONET OC-3c. Because it is a wireless solution, it requires a minimal infrastructure and can be deployed quickly and cost-effectively as an alternative to the wired infrastructure provided by incumbent service providers. After all, the highest cost component (more than 50 percent) when building networks is not the distribution facility, but rather the labor required to trench it into the ground or build aerial facilities. Thus, any access alternative that minimizes the cost of labor will garner significant attention.

LMDS relies on a cellular-like deployment strategy under which the cells are approximately three miles in diameter. Unlike cellular service, however, users are stationary. Consequently, there is no need for LMDS cells to support roaming. Antenna/transceiver units are generally placed on rooftops as they need unobstructed line of sight to operate properly. In fact, this is one of the disadvantages of LMDS (and a number of other wireless technologies). Besides suffering from unexpected physical obstructions, the service suffers from *rain fade* caused by the absorption and scattering of the transmitted microwave signal by atmospheric moisture. Even some forms of foliage will cause interference for LMDS, so the transmission and reception equipment must be mounted high enough to avoid such obstacles—hence, the tendency to mount the equipment on rooftops.

Because of its high-bandwidth capability, many LMDS implementations interface directly with an ATM backbone to take advantage of both its bandwidth and its diverse QoS capability. If ATM is indeed the transport fabric of choice, then the LMDS service becomes a broadband access alternative to a network capable of transporting a full range of services including voice, video, image, and data—the full suite of multimedia applications.

Multichannel Multipoint Distribution System (MMDS)
MMDS got its start as a "wireless cable television" solution. In 1963, a spectrum allocation known as the *Instructional Television Fixed Service* (ITFS) was carried out by the FCC as a way to distribute educational content to schools and universities. In the 1970s, the FCC established a two-channel metropolitan distribution service called the *Multipoint Distribution Service* (MDS). It was to be used for the delivery of pay-TV signals, but with the advent of inexpensive satellite access and the ubiquitous deployment of cable systems, the need for MDS went away.

In 1983, the FCC rearranged the MDS and ITFS spectrum allocation, creating 20 ITFS education channels and 13 MDS channels. In order to qualify to use the ITFS channels, schools had to use a minimum of 20 hours of airtime, which meant that ITFS channels tended to be heavily, albeit randomly, utilized. As a result, MMDS providers that use all 33 MDSs and ITFSs must be able to dynamically map requests for service to available channels in a completely transparent fashion, which means that the bandwidth management system must be reasonably sophisticated.

Because MMDS is not a true cable system (in spite of the fact that it has its roots in television distribution), there are no franchise issues for its use (there are, of course, licensing requirements). However, the technology is also limited in terms of what it can do. Unlike LMDS, MMDS is designed as a one-way broadcast technology and therefore does not typically allow for upstream communication. Many contend, however, that there is adequate bandwidth in MMDS to provision two-way systems, which would make it suitable for voice, Internet access, and other data-oriented services.

Satellite Technology In October 1945, Arthur C. Clarke published a paper in *Wireless World* entitled "Extra-Terrestrial-Relays: Can Rocket Stations Give World-Wide Radio Coverage?" In his paper, Clarke proposed the concept of an orbiting platform that would serve as a relay facility for radio signals sent to it that could be turned around and retransmitted back to the earth with far greater coverage than was achievable through terrestrial transmission techniques. His platform would orbit at an altitude of 42,000 kilometers (25,200 miles) above the equator where it would orbit at a speed identical to the rotation speed of the earth. As a consequence, the satellite would appear to be stationary to earth-bound users.

Satellite technology may prove to be the primary communications gateway for regions of the world that do not yet have a terrestrial wired infrastructure, particularly given the fact that they are now capable of delivering broadband services. In addition to the United States, the largest markets for satellite coverage are Latin America and Asia, particularly Brazil and China.

Geosynchronous Satellites   Clarke's concept of a stationary platform in space forms the basis for today's geostationary or geosynchronous satellites. Ringing the equator like a string of pearls, these devices provide a variety of services including 64 Kbps voice, broadcast television, *video on demand* (VOD) services, broadcast and interactive data, and point-of-sale applications, to name a few. Although satellites are viewed as technological marvels, the real magic lies more with what it takes to harden them for the environment in which they must operate and what it takes to get them there than it does their actual operational responsibilities. Satellites are, in effect, nothing more than a sophisticated collection of assignable, on-demand repeaters—in a sense, the world's longest local loop.

From a broadcast perspective, satellite technology has a number of advantages. First, its one-to-many capabilities are unequaled. Information from a central point can be transmitted to a satellite in geostationary orbit; the satellite can then rebroadcast the signal back to the earth, covering an enormous service footprint.

Because the satellites appear to be stationary, the earth stations actually *can* be. One of the most common implementations of geosynchronous technology is seen in the *Very Small Aperture Terminal* (VSAT) dishes that have sprung up like mushrooms on a summer lawn. These dishes are used to provide both broadcast and interactive applications. The small *direct broadcast satellite* (DBS) dishes used to receive TV signals are examples of broadcast applications, while the dishes seen on the roofs of large retail establishments, automobile dealerships, and convenience stores are typically (although not always) used for interactive applications such as credit-card verification, inventory queries, e-mail, and other corporate communications. Some of these applications use a satellite downlink, but rely on a telco return for the upstream traffic—that is, they must make a telephone call over a landline to offer two-way service.

One disadvantage of geosynchronous satellites has to do with their orbital altitude. On one hand, because they are so high, their service footprint is extremely large. On the other hand, because of the distance from the earth to the satellite, the typical transit time for the signal to propagate from the ground to the satellite (or back) is about half a second, which is a significant propagation delay for many services. Should an error occur in the transmission stream during transmission, the need to detect the error, ask for a retransmission, and wait for the second copy to arrive could be catastrophic for delay-sensitive services like voice and video. Consequently, many of these systems rely on forward error correction transmission techniques that enable the receiver to not only detect the error, but correct it as well.

An interesting observation is that because the satellites orbit above the equator, dishes in the northern hemisphere always face south. The farther north a user's receiver dish is located, the lower it has to be oriented. Where I live in Vermont, the satellite dishes are practically lying on the ground—they almost look as of they are receiving signals from the depths of the mountains instead of a satellite orbiting 23,000 miles above the earth.

Low/Medium Earth Orbit Satellites (LEO/MEO) In addition to the geosynchronous satellite arrays, there are a variety of lower orbit constellations deployed known as *Low / Medium Earth Orbit* (LEO/MEO) satellites. Unlike the geosynchronous satellites, these orbit at lower altitudes—400 to 600 miles, which is much lower than the 23,000-mile altitude of the typical GEO bird. As a result of their lower altitude, the transit delay between an earth station and a LEO satellite is virtually nonexistent. However, another problem exists with LEO technology. Because the satellites orbit pole to pole, they do not appear to be stationary, which means that if they are to provide uninterrupted service, they must be able to hand off a transmission from one satellite to another before the first bird disappears below the horizon. This has resulted in the development of sophisticated satellite-based technology that emulates the functionality of a cellular telephone network. The difference is that in this case, the user does not appear to move—the cell does!

As a service-provisioning technology, satellites may seem so far out (no pun intended) that they may not appear to pose a threat to more traditional telecommunications solutions. At one time, that is, before the advent of LEO technology, this was largely true. Geosynchronous satellites were extremely expensive, offered low bit rates, and suffered from serious latency that was unacceptable for many applications.

This is no longer true. Today, GEO satellites offer high-quality, two-way transmission for certain applications. LEO technology has advanced to the point that it now offers low-latency, two-way communications at broadband speeds, is relatively inexpensive, and as a consequence poses a clear threat to terrestrial services. On the other hand, the best way to eliminate an enemy is to make the enemy a friend. Many traditional service providers have entered into alliances with satellite providers; consider the agreements that exist among DirecTV (a high-quality, wireless alternative to cable), Bell Atlantic, GTE, Cincinnati Bell, and SBC corporations. By joining forces with satellite providers, service providers create a market block that will help them stave off the short-term incursion of cable. Between the minimal infrastructure required to receive satellite signals and the soon-to-be ubiquitous deployment of DSL over twisted pair, incumbent local telephone companies and their alliance partners are in a reasonably good position to counter the efforts of cable providers wanting to enter the local services marketplace. In the long term, however, wireless will win the access game.

Other Wireless Access Solutions

A number of other wireless technologies have emerged in the last few years that are worth mentioning, including 802.11, Bluetooth, and the *Wireless Application Protocol* (WAP).

802.11 IEEE 802.11 is a wireless LAN standard developed by the IEEE's 802 committee to specify an air interface between a wireless client and a base station, as well as among a variety of wireless clients. First discussed in 1990, the standard has evolved through six draft versions and won final approval on June 26, 1997.

802.11 PHY Layer All 802 standards address themselves to both the *physical* (PHY) and *Media Access Control* (MAC) layers. At the PHY layer, IEEE 802.11 identifies three options for wireless LANs: diffused infrared, *direct sequence spread spectrum* (DSSS), and *frequency hopping spread spectrum* (FHSS).

While the infrared PHY operates at a baseband level, the other two radios operate at 2.4 GHz, part of the *Industrial, Scientific, and Medical* (ISM) band. It can be used for operating wireless LAN devices and does not require an end-user license. All three PHYs specify support for 1 Mbps and 2 Mbps data rates.

802.11 MAC Layer The 802.11 MAC layer, like CSMA/CD and token passing, presents the rules used to access the wireless medium. The primary services provided by the MAC layer are as follows:

- *Data transfer* Based on a CSMA/CA algorithm as the media access scheme.
- *Association* The establishment of wireless links between wireless clients and *access points* (APs).
- *Authentication* The process of conclusively verifying a client's identity prior to a wireless client associating with an AP. 802.11 devices operate under an open system where any wireless client can associate with any AP without verifying credentials. True authentication is possible with the use of the *Wired Equivalent Privacy Protocol* (WEP), which uses a shared key validation protocol similar to that used in *Public Key Infrastructure* (PKI).

Only those devices with a valid shared key can be associated with an AP.

- *Privacy* By default, data is transferred "in the clear;" any 802.11-compliant device can potentially eavesdrop PHY 802.11 traffic that is within range. WEP encrypts the data before it is transmitted using a 40-bit encryption algorithm known as RC4. The same shared key used in authentication is used to encrypt or decrypt the data. Only clients with the correct shared key can decipher the data.

- *Power management* 802.11 defines an *active mode*, where a wireless client is powered at a level adequate to transmit and receive, and a *power-save mode*, under which a client is not able to transmit or receive, but consumes less power while in a standby mode of sorts.

802.11 has garnered a great deal of attention in recent months, particularly with the perceived competition from Bluetooth, another short-distance wireless protocol. However, significantly more activity is underway in the 802.11 space with daily product announcements throughout the industry. Various subcommittees have been created that address everything from security to voice transport to QoS. It is a technology to watch.

Bluetooth Bluetooth has been referred to as the *personal area network* (PAN). It is a wireless LAN on a chip that operates in the unlicensed 2.4 GHz band at 768 Kbps, which is relatively slow compared to 802.11's 11 Mbps. It does, however, include a 56 Kbps backward channel, three voice channels, and can operate at distances of up to 100 feet (although most pundits claim 25 feet for effective operation). According to a report from Allied Business Intelligence, the Bluetooth devices market will reach $2 billion by 2005, a nontrivial number.

The service model that Bluetooth supporters propose is one built on the concept of the mobile appliance. Consider the following scenario. As you walk around your house with your Palm Pilot or Pocket PC, the device is in constant communication with Bluetooth-equipped devices throughout the house. As you pass by the refrigerator on the way through the kitchen, the fridge transmits a message to your device telling it that the milk is low and should be added to the shopping list. It knows this because infrared sensors inside the refrigerator have detected that the level of milk in the container is below a certain predetermined level. The mobile appliance adds milk to the shopping list in the mobile device. The next time you are out and pass the grocery store, the mobile appliance receives a transmission from the store, wakes up, and notifies you to stop and buy the items on the list. Good application? Maybe. Bluetooth, named after a tenth-century Danish king, is experiencing growing pains and significant competition from 802.11 for many good reasons. Whether Bluetooth succeeds or not is a matter still open for discussion. It's much too early to tell.

Wireless Application Protocol (WAP)   Originally developed by Phone.com, WAP has proven to be a disappointment for the most part. Because it is designed to work with 3G wireless systems, and because 3G systems have not yet materialized, some have taken to defining WAP to mean "Wrong Approach to Portability." Germany's D2 network reports that the average WAP customer uses it less than two minutes per day, which is tough to make money on when service is billed on a usage basis. 3G will be the deciding factor; when it succeeds, WAP will succeed—unless 802.11's success continues to expand.

The Mobile Appliance

The mobile appliance concept is enjoying a significant amount of recent attention because it promises to herald in a whole new way of using network and computer resources—*if it works as promised*. The problem with so many of these new technologies is that they overpromise and underdeliver—precisely the opposite of what they're supposed to do for a successful rollout. 3G, for example, has been billed as "the wireless Internet." It has failed largely as a result of that billing. It is *not* the Internet—it is far from it. The bandwidth isn't there, nor is a device that can even begin to offer the kind of image quality that Internet users have become accustomed to. Furthermore, the number of screens that a user must go through to reach a desired site (I have heard estimates as high as 22!) is far too high. Therefore, until the user interface, content, and bandwidth challenges are met and satisfied, the technology will remain exactly that—a technology. There is no application yet, and *that's* what people are willing to pay money for.

Wireless Access Summary

Access technologies, which are used to connect the customer to the network, come in a variety of forms and offer a broad variety of connectivity options and bandwidth levels. The key to success is to *not* be a bottleneck. Access technologies that can evolve to meet the growing customer demands for bandwidth will be the winners in the game. DSL holds an advantage as long as it can overcome the availability challenge and the technology challenge of loop carrier restrictions. Wireless is hobbled by licensing and spectrum availability, both of which are regulatory and legal in nature rather than technology limitations.

We have now discussed the premises environment and the access technologies that connect them to the wider-area network. Next, we'll examine transport technologies.

Transport Technologies

Because businesses are rarely housed in a single building and because customers are typically scattered across a broad geographical area (particularly multinational customers), there is a growing need for high-speed, reliable wide area transport. "Wide area" can take on a variety of meanings. For example, a company with multiple offices scattered across the metropolitan expanse of a large city requires interoffice connectivity in order to do business properly. On the other hand, a large multinational company with offices and clients in Madrid, San Francisco, Hamburg, and Singapore requires connectivity to ensure that the offices can exchange information on a 24-hour basis.

These requirements are satisfied through the proper deployment of wide area transport technologies. These can be as simple as a dedicated private-line circuit or as complex as a virtual installation that relies on ATM for high-quality transport.

Dedicated facilities are excellent solutions because they are dedicated. They provide fixed bandwidth that never varies and guarantee the quality of the transmission service. Because they are dedicated, however, they suffer from two disadvantages. First, they are expensive and only cost-effective when highly utilized. The pricing model for dedicated circuits includes two components: the mileage of the circuit and the bandwidth. The longer the circuit and the faster it is, the more it costs. Second, because they are not switched and are often not redundant because of cost, dedicated facilities pose the potential threat of a prolonged service outage should they fail. Nevertheless, dedicated circuits are popular for certain applications and widely deployed. They include such solutions as T1, which offers 1.544 Mbps of bandwidth; DS3, which offers 44.736 Mbps of bandwidth; and SONET, which offers a wide range of bandwidth from 51.84 Mbps to as much as 40 Gbps.

The alternative to a dedicated facility is a switched service, such as frame relay or ATM. These technologies provide *virtual circuits*—in other words, instead of dedicating physical facilities, they dedicate logical time slots to each customer who then shares access to physical network resources. In the case of frame relay, the service can provide bandwidth as high as DS3, thus providing an ideal replacement technology for lower-speed dedicated circuits. ATM, on the other hand, operates hand-in-glove with SONET and is thus capable of providing transport services at gigabit speeds. Finally, the new field of optical networking is carving out a large niche for itself as a bandwidth-rich solution with the potential for inherent QoS.

We begin our discussion with dedicated private line, otherwise known as point-to-point.

Point-to-Point Technologies

Point-to-point technologies do exactly what their name implies: They connect one point directly with another. For example, it is common for two buildings in a downtown area to be connected by a point-to-point microwave or infrared circuit because the cost of establishing it is much lower than the cost to put in physical facilities in a crowded city. Many businesses rely on dedicated, point-to-point optical facilities to interconnect locations, especially businesses that require dedicated bandwidth for high-speed applications. Of course, point-to-point does not necessarily imply high bandwidth; many locations use 1.544 Mbps T1 facilities for interconnection and some rely on lower-speed circuits where higher bandwidth is not required.

Dedicated facilities provide bandwidth from as low as 2,400 bits per second to as high as multiple gigabits per second. 2,400 bps analog facilities are not commonly seen, but are often used for alarm circuits and telemetry, whereas circuits operating at 4,800 and 9,600 bps are used to access interactive, host-based data applications.

Higher-speed facilities are usually digital and are often channelized by dedicated multiplexers and shared among a collection of users or by a variety of applications. For example, a high-bandwidth facility that interconnects two corporate locations might be dynamically subdivided into various-sized channels for use by a PBX for voice, a videoconferencing system, and data traffic.

Dedicated facilities have the advantage of always being available to the subscriber. However, they have the disadvantage of being there and accumulating charges whether or not they are being used. For the longest time, dedicated circuits represented the only solution that provided guaranteed bandwidth—switched solutions simply weren't designed for the heavy service requirements of graphical and data-intensive traffic. Over time, however, that has changed. A number of switched solutions have emerged in the last few years that provide guaranteed bandwidth and only accumulate charges when they are being used (although some of them offer very reasonable fixed-rate service). The two most common of these are frame relay and ATM. Before we discuss these, however, let's spend a few minutes discussing the hierarchy of switching. Part of this is review of prior material; most of it, though, is preparation for our discussion of frame relay and ATM—the so-called fast packet switching technologies. We begin our discussion with point-to-point technologies.

Dedicated Services: T1 and E1   The original network was exclusively analog until 1962 when T-carrier emerged to provide connectivity between central offices. The technology was originally introduced as a short-haul, 4-wire facility that was designed to serve metropolitan areas, a technology that customers would never have a reason to know about—after all, what customer could ever need 1.544 Mbps of bandwidth?

Consider the following scenario. A corporation designs and builds a new videoconferencing facility in their headquarters building. The facility will be used for high-end, high-impact conferencing and must therefore have access to high-bandwidth facilities.

The telephone company has several options that it can use to provide access connectivity. It can bring in a series of ISDN BRI lines or it can provision the necessary bandwidth through digital carrier systems that transport multiple circuits over a single shared facility. This second option is clearly the most cost-effective and is the option that is most commonly used for these kinds of installations because it enables the customer to provision bandwidth on demand as required by the application.

The most common form of multiplexed access and transport is T-carrier, or E-carrier outside the United States. Let's take a few minutes to describe them.

Framing and Formatting in T1   The standard T-carrier multiplexer, shown in Figure 1-51, has 24 channels, each capable of delivering 64 Kbps of bandwidth. The channels can be clustered in a variety of ways to provide a customer with increasing amounts of bandwidth, from a single 64 channel up to the entire 1,536 Mbps of bandwidth that is inherently available in the carrier system. If the videoconference unit requires 384 Kbps for optimal performance, then six 64 Kbps channels are bonded together to create the higher-bit-rate channel.

The Rest of the World: E1   E1, used for the most part outside of the United States and Canada, differs from T1 on several key points. First, it boasts a 2.048 Mbps facility rather than the 1.544 Mbps facility found in T1. Second, it utilizes a 32-channel frame rather than 24. Channel 1 contains framing information and a 4-bit *cyclic*

redundancy check (CRC-4), channel 16 contains all signaling information for the frame, and channels 1 through 15 and 17 through 31 transport user traffic. As with T1, channels can be bonded together to form a superrate channel for video applications.

Here's an interesting point about T1 and E1: because T1 is a departure from the international E1 standard, it is incumbent upon the T1 provider to perform all interconnection conversions between T1 and E1 systems. For example, if a call arrives in the United States from a European country, the receiving American carrier must convert the incoming E1 signal to T1. If a call originates from Canada and is terminated in Australia, the originating Canadian carrier must convert the call to E1 before transmitting it to Australia.

Beyond T1 When T1 and E1 first emerged on the telecommunications scene, they represented a dramatic step forward in terms of the bandwidth that service providers now had the ability to provide. In short order, however, as increasing requirements for bandwidth drove incessant demand, broadband transmission systems like T1 became mainstream, usage went up, and soon requirements emerged for digital transmission systems with capacity greater than 1.544 Mbps. Before long, we had created the *North American Digital Hierarchy* with an upward bandwidth limit of 45 Mbps (DS3). Although many believed that DS3 was the maximum that anyone would ever require in the way of transport bandwidth, others felt that we were just beginning to plumb the depths of possibility with regard to the applications that were being created. The next great bandwidth leap was optical, and it has become critically important in many applications, particularly bandwidth-intensive applications such as video, audio, and other multimedia sources. The solution is called SONET in North America and SDH in the rest of the world, and it dramatically simplifies the world of high-speed transport.

New Services   SONET bandwidth is imminently scalable, meaning that the ability to provision additional bandwidth for customers that require it on an as-needed basis becomes real. As applications evolve to incorporate more multimedia content and therefore require greater volumes of bandwidth, SONET offers it by the bucketload. Interfaces between SONET and Gigabit Ethernet are already being written; interfaces to ATM and other high-speed switching architectures have been in existence for some time.

SONET was initially designed to provide multiplexed point-to-point transport. However, as its capabilities became better understood and networks became mission critical, its deployment became more innovative and soon it was deployed in ring architectures, as shown in Figure 1-52. These rings represent one of the most commonly deployed network topologies. For the moment, however, let's examine a point-to-point deployment. As it turns out, rings don't differ all that much.

If we consider the structure and function of the typical point-to-point circuit, we find a variety of devices and functional regions, as shown in Figure 1-53. In this case, the components include end devices; multiplexers, which provide the point of entry for traffic originating in the customer's equipment and seeking transport across the network; a full-duplex circuit, which provides simultaneous two-way transmission between the network components; a series of repeaters or regenerators, which are responsible for periodically

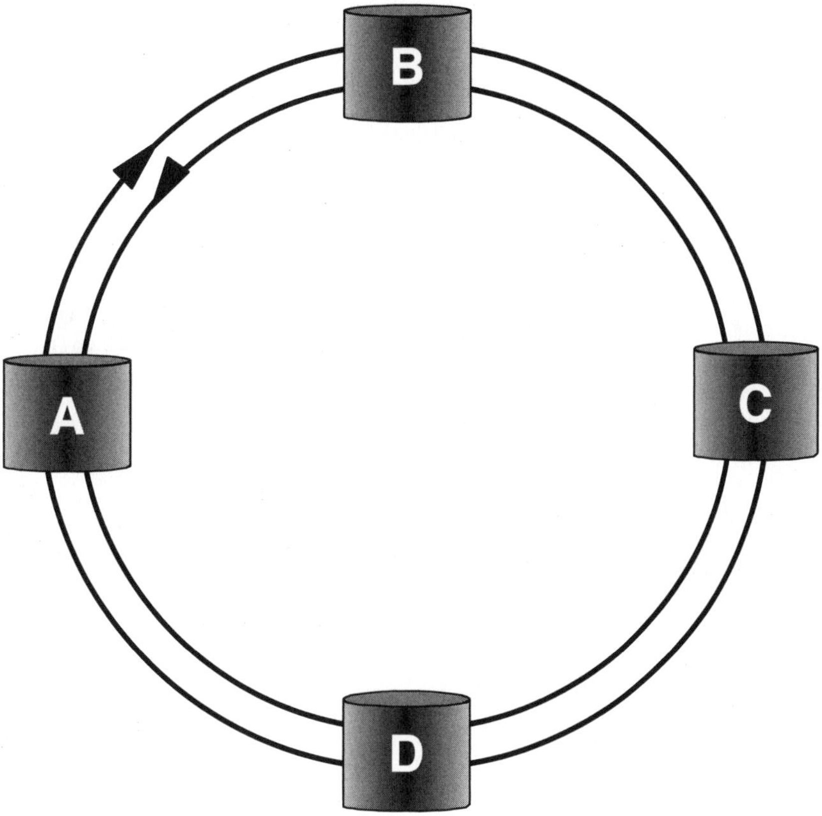

Figure 1-52
Ring architecture

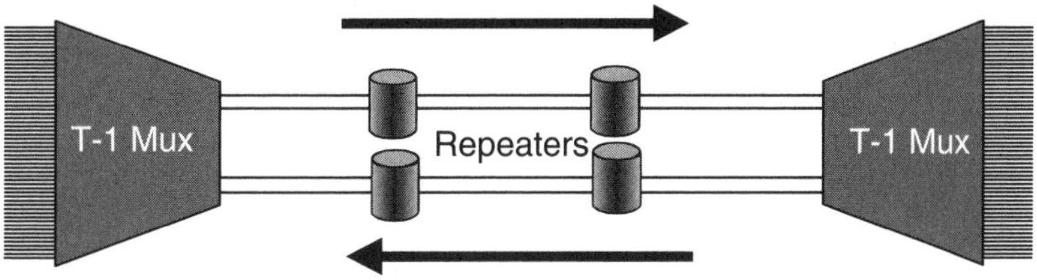

Figure 1-53
Architecture of a typical T-carrier span

reframing and regenerating the digital signal; and one or more intermediate multiplexers, which serve as nothing more than pass-through devices.

When non-SONET traffic is transmitted into a SONET network, it is packaged for transport through a step-by-step, quasi-hierarchical process that attempts to make reasonably good use of the available network bandwidth and ensure that receiving devices can interpret the data when it arrives. The intermediate devices, including multiplexers and repeaters, also play a role in guaranteeing traffic integrity, and to that end, the SONET standards divide the network into three regions: path, line, and section. To understand the differences between the three, let's follow a typical transmission of a DS3, probably carrying 28 T1s, from its origination point to the destination.

When the DS3 first enters the network, the ingress SONET multiplexer packages it by wrapping it in a collection of additional information called *path overhead*, which is unique to the transported data. For example, it attaches information that identifies the original source of the DS3 so that it can be traced in the event of network transmission problems, a bit-error control byte, information about how the DS3 is actually mapped into the payload transport area (and unique to the payload type), an area for network performance and management information, and a number of other informational components that have to do with the end-to-end transmission of the unit of data.

The packaged information, now known as a *payload*, is inserted into a SONET frame, and at that point another layer of control and management information is added called *line overhead*. Line overhead is responsible for managing the movement of the payload from multiplexer to multiplexer. To do this, it adds a set of bytes that enables receiving devices to find the payload inside the SONET frame. As you will learn a bit later, the payload can occasionally wander around inside the frame due to the vagaries of the network. These bytes enable the system to track that movement.

In addition to these tracking bytes, the line overhead includes bytes that monitor the integrity of the network and have the capability to affect a switch to a backup transmission span if a failure in the primary span occurs. It also includes another bit-error-checking byte, a robust channel for transporting network management information, and a voice communications channel that enables technicians at either end of the line to plug in with a handset and communicate while troubleshooting.

The final step in the process is to add a layer of overhead that enables the intermediate repeaters to find the beginning of and synchronize a received frame. This overhead, called the *section over-head*, contains a unique initial framing pattern at the beginning of the frame, an identifier for the payload signal being carried, another bit-error check, a voice communications channel, and another dedicated channel for network management information, which is similar to but smaller than the one identified in the line overhead.

The result of all this overhead, much of which seems like overkill (and for many is overkill), is that the transmission of a SONET frame containing user data can be identified and managed with tremendous granularity from the source all the way to the destination.

So, to summarize, the hard little kernel of DS3 traffic is gradually surrounded by three layers of overhead information, as shown in Figure 1-54, that help it achieve its goal of successfully making its way across the network. The section overhead is used at every device the signal passes through, including multiplexers and repeaters. The line overhead is only used between multiplexers. The information contained in the path overhead is only used by the source and destination multiplexers—the intermediate multiplexers don't care about the specific nature of the payload because they don't have to terminate or interpret it.

Figure 1-54
Layers of SONET overhead surround the payload prior to transmission.

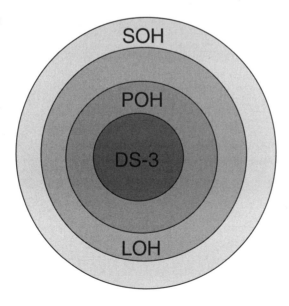

The SONET Frame   Keep in mind once again that we are doing nothing more complicated than building a T1 frame with an attitude. Recall that the T1 frame comprised 24 8-bit channels (samples from each of 24 incoming data streams) plus a single bit of overhead. In SONET, we have a similar construct—much more channel capacity and much more overhead, but the same functional concept.

The fundamental SONET frame is shown in Figure 1-55 and is known as a *Synchronous Transport Signal, Level One* (STS-1). It is 9 bytes tall and 90 bytes wide, for a total of 810 bytes of transported data including both user payload and overhead. The first three columns of the frame are the section and line overhead, which are known collectively as the *transport overhead*. The bulk of the frame itself, to the left, is the *synchronous payload envelope* (SPE), which is the container area for the user data that is being transported. The data, previously identified as the payload, begins somewhere in the payload envelope. The path overhead begins when the payload begins. Because it is unique to the payload itself, it travels closely with the payload. The first byte of the payload is actually the first byte of the path overhead.

The SONET frame is transmitted serially on a row-by-row basis. The SONET multiplexer transmits (and therefore receives) the first byte of row one all the way to the 90th byte of row one, and then

Figure 1-55
Fundamental
SONET frame

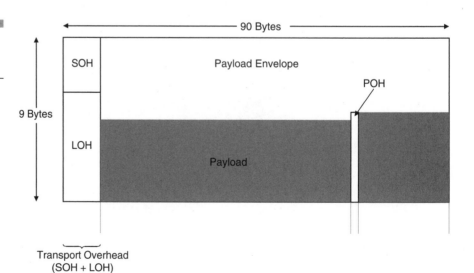

wraps to transmit the first byte of row two all the way to the 90th byte of row two, and so on until all 810 bytes have been transmitted.

Because the rows are transmitted serially, the many overhead bytes do not all appear at the beginning of the transmission of the frame—instead, they are peppered along the bit stream like highway markers. For example, the first 2 bytes of overhead in the section overhead are the framing bytes followed by the single-byte signal identifier. The next 87 bytes are user payload followed by the next byte of section overhead—in other words, there are 87 bytes of user data between the first 3 section overhead bytes and the next one! The designers of SONET were clearly thinking the day they came up with this because each byte of data appears just when it is needed. Truly a remarkable thing!

Because of the unique way that the user's data is mapped into the SONET frame, the data can actually start pretty much anywhere in the payload envelope. The payload is always the same number of bytes, which means that if it starts late in the payload envelope, it may run into the payload envelope of the next frame! In fact, this happens more often than not, but it's okay—SONET is equipped to handle this odd behavior.

SONET Bandwidth The SONET frame consists of 810 8-bit bytes and, like the T1 frame, is transmitted once every 125 μsec (8,000 frames per second). Doing the math, this works out to an overall bit rate of

$$810 \text{ bytes/frame} \times 8 \text{ bits/byte} \times 8,000 \text{ frames/second} = 51.84 \text{ Mbps}$$

which is the fundamental transmission rate of the SONET STS-1 frame.

That's a lot of bandwidth—51.84 Mbps is slightly more than a 44.736 Mbps DS3, a respectable carrier level by anyone's standard. What if more bandwidth is required, however? What if the user wants to transmit multiple DS3s or perhaps a single signal that requires more than 51.84 Mbps, such as a 100 Mbps Fast Ethernet signal or a *high-definition television* (HDTV) transmission? Or for that matter, what about a payload that requires less than 51.84 Mbps? In those cases, we have to invoke more of SONET's magic.

The STS-N Frame   In situations where multiple STS-1s are required to transport multiple payloads, all of which fit in an STS-1's payload capacity, SONET allows for the creation of *STS-N frames*, where *N* represents the number of STS-1 frames that are multiplexed together to create the frame. If three STS-1s are combined, the result is an STS-3. In this case, the three STS-1s are brought into the multiplexer and *byte interleaved* to create the STS-3, as shown in Figure 1-56. In other words, the multiplexer selects the first byte of frame one, followed by the first byte of frame two, followed by the first byte of frame three. Then it selects the second byte of frame one, followed by the second byte of frame two, followed by the second byte of frame three, and so on until it has built an interleaved frame that is now three times the size of an STS-1: 9×270 bytes instead of 9×90. Interestingly (and impressively), the STS-3 is still generated 8,000 times per second.

The technique described previously is called a *single stage multiplexing process* because the incoming payload components are combined in a single step. There is also a two-stage technique that is commonly used. For example, an STS-12 can be created in two ways. Twelve STS-1s can be combined in a single stage process to create

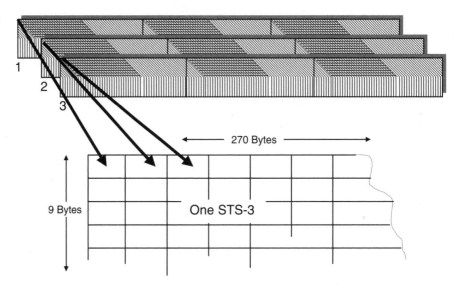

Figure 1-56
Byte interleaving in SONET

the byte-interleaved STS-12. Alternatively, four groups of three STS-1s can be combined to form four STS-3s, which can then be further combined in a second stage to create a single STS-12. Obviously, two-stage multiplexing is more complex than its single-stage cousin, but both can be used.

NOTE: *The overall bit rate of the STS-N system is N × STS-1. However, the maximum bandwidth that can be transported is STS-1, but N of them can be transported! This is analogous to a channelized T1.*

The STS-Nc Frame Let's go back to our Fast Ethernet example mentioned earlier. In this case, 51.84 Mbps is inadequate for our purposes because we have to transport the 100 Mbps Ethernet signal. For this, we need a *concatenated signal*.

Concatenate means "to string together," which is exactly what we do when we need to create a *superrate frame*—in other words, a frame capable of transporting a payload that requires more bandwidth than an STS-1 can provide, such as our 100 Mbps Fast Ethernet frame. In the same way that an STS-N is analogous to a channelized T1, an STS-Nc is analogous to an unchannelized T1. In both cases, the customer is given the full bandwidth that the pipe provides; the difference lies in how the bandwidth is parceled out to the user.

Transporting Subrate Payloads: Virtual Tributaries (VTs) When a SONET frame is modified for the transport of subrate payloads, it is said to carry *virtual tributaries* (VTs). Simply put, the payload envelope is chopped into smaller pieces that can then be individually used for the transport of multiple lower-bandwidth signals.

Creating VTs To create a VT-ready STS, the synchronous payload envelope is subdivided. An STS-1 comprises 90 columns of bytes, 4 of which are reserved for overhead functions (section, line, and path). That leaves 86 for actual user payload. To create VTs, the payload capacity of the SPE is divided into 7 12-column pieces called *virtual tributary groups*. Math majors will be quick to point out that 7 × 12

= 84, leaving 2 unassigned columns. These columns, shown in Figure 1-57, are indeed unassigned and are given the rather silly name of *fixed stuff.*

Now comes the fun part. Each of the VT groups can be further subdivided into one of four different VTs to carry a variety of payload types, as shown in Figure 1-58. A VT1.5, for example, can easily transport a 1.544 Mbps signal within its 1.728 Mbps capacity, with a little room left over. A VT2, meanwhile, has enough capacity in its 2.304 Mbps structure to carry a 2.048 Mbps European E1 signal, with a little room left over. A VT3 can transport a DS-1C signal, while a VT6 can easily accommodate a DS2, again, each with a little room left over.

One aspect of VTs that must be mentioned is the mix-and-match nature of the payload. Within a single SPE, the seven VT groups can carry a variety of different VTs. However, each VT group can carry only one VT type.

The previous "little room left over" comment is, by the way, one of the key points that SONET and SDH detractors point to when criticizing them as legacy technologies, claiming that in these times of growing competition and the universal drive for efficiency, they are inordinately wasteful of bandwidth, given that they were designed when the companies that delivered them were monopolies and less

Figure 1-57

SONET fixed stuff

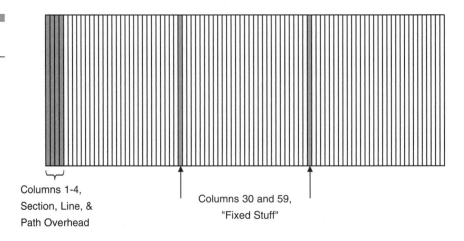

Columns 1-4,
Section, Line, &
Path Overhead

Columns 30 and 59,
"Fixed Stuff"

Figure 1-58
SONET VT payload
transport

VT Type	Columns/ VT	Bytes/VT	VTs/Group	VTs/SPE	VT Bandwidth
VT1.5	3	27	4	28	1.728
VT2	4	36	3	21	2.304
VT3	6	54	2	14	3.456
VT6	12	108	1	7	6.912

concerned about such things than they are now. Suffice it to say that one of the elegant aspects of SONET is its capability to accept essentially any form of data signal, map it into standardized positions within the SPE frame, and transport it efficiently and at a very high speed to a receiving device on the other side of town or the other side of the world.

SONET Summary SONET is a complex and highly capable standard designed to provide high-bandwidth transport for legacy and new protocol types alike. The overhead that it provisions has the capability to deliver a remarkable collection of network management, monitoring, and transport granularity.

The European SDH shares many of SONET's characteristics, as we will now see. SONET, you will recall, is a limited North American standard for the most part. SDH, on the other hand, provides high-bandwidth transport for the rest of the world.

Most books on SONET and SDH cite a common list of reasons for their proliferation, including a recognition of the importance of the global marketplace and a desire on the part of manufacturers, therefore, to provide devices that will operate in both SONET and SDH environments, the global expansion of ring architectures, a greater focus on network management and the value that it brings to the table, and massive, unstoppable demand for more bandwidth. To those features, add these: an increasing demand for high-speed routing capability to work hand-in-glove with transport; the deployment of DS1, DS3, and E1 interfaces directly to the enterprise customer as access solutions; growth in the demand for broadband access technologies such as cable modems, the many flavors of DSL, and two-way satellite connectivity; the ongoing replacement of traditional circuit-switched network fabrics with packet-based transport and mesh architectures; a renewed focus on the SONET and SDH overhead with an eye toward using it more effectively; and the convergence of multiple applications on a single, capable, high-speed network fabric. The most visible among these is the hunger for bandwidth. According to the consultant group RHK, global volume demand will grow from approximately 350,000 terabytes of transported data per month in April 2000 to more than 16 million terabytes of traffic per month in 2003. And who can argue?

SONET and SDH were originally rolled out to replace the T1 and E1 hierarchies, which were suffering from demands for bandwidth beyond what they were capable of delivering. Their principle deliverable was voice—and lots of it. However, as time passed and data made its presence known, SONET and SDH found use in a wide variety of network applications. Video is clearly one of these applications, particularly high-end video. Although SONET or SDH may not be required for the typical videoconferencing system installed by most corporations, from a service provider's perspective, these technologies are ideal as future-proof solutions to the growing demands for broadband that multimedia applications are placing on the network.

The Switching Hierarchy

The switching hierarchy, shown in Figure 1-59, has two major sub-headings—circuit switching and store-and-forward switching. Circuit switching is something of an evolutionary dead end in that it will not become something else. The only evolution for circuit switching is an evolution toward packet switching at this point.

Packet switching evolved as one of the two descendents of store-and-forward technology. Message switching, the alternative and another evolutionary dead end, is inefficient and not suited to the bursty nature of most data services today. Packet switching continues to hold sway and because of its many forms, it is a valid solution for most data applications.

Packet switching has three major forms, two of which were discussed in earlier chapters. Connection-oriented packet switching manifests itself as virtual circuit service that offers the appearance and behavior of a dedicated circuit when in fact it is a switched service. It creates a path through the network that all packets from the same source (and going to the same destination) follow, the result of

Figure 1-59
The switching
hierarchy

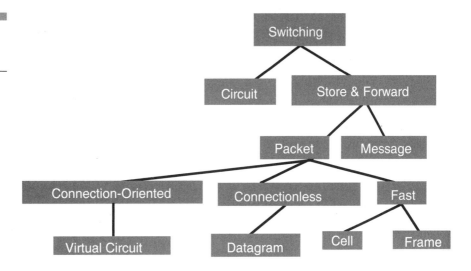

which is constant transmission latency for all packets in the flow. Both frame relay and ATM, discussed later in the section "The Switching Heirarchy," are virtual circuit technologies.

Connectionless packet switching does not establish a virtual dedicated path through the network; instead, it hands the packets to the ingress switch and enables the switch to determine optimum routing on a packet-by-packet basis. This is extremely efficient from a network standpoint, but is less favorable for the client in that every packet is treated slightly differently, the result of which can be unpredictable delay and out-of-order delivery. IP is an example of a connectionless protocol.

The third form of packet switching is called *fast packet*, two forms of which are frame relay and ATM. The technique is called fast packet because the processing efficiency is far better than traditional packet switching because of reduced overhead.

Fast packet technologies are characterized by low error rates, significantly lower processing overhead, high transport speed, minimal delay, and relatively low cost. The switches accomplish this by making several assumptions about the network. First, they assume (correctly) that the network is digital and based largely on a fiber infrastructure, the result of which is an inordinately low error rate. Second, they assume that unlike their predecessors, the end-user devices are intelligent and therefore have the capability to detect and correct errors on an end-to-end basis (at a higher protocol layer) rather than stopping every packet at every node to check it for errors and correct them. These switches still check for errors, but if they find errored packets, they simply discard them, knowing that the end devices will realize that there is a problem and take corrective measures on an end-to-end basis.

Frame Relay Frame relay came about as a private-line replacement technology and was originally intended as a data-only service. Today, it carries not only data, but voice and video as well, and although it was originally crafted with a top speed of T1/E1, it now provides connectivity at much higher-bandwidth levels.

In frame relay networks, the incoming data stream is packaged as a series of variable-length frames that can transport any kind of data—LAN traffic, IP packets, SNA frames, and even voice and video. In fact, it has been recognized as a highly effective transport mechanism for voice, enabling frame-relay-capable PBXs to be connected to a frame relay *permanent virtual circuit* (PVC), which can cost-effectively replace private-line circuits used for the same purpose. When voice is carried over frame relay, it is usually compressed for transport efficiency and packaged in small frames to minimize the processing delay of the frames. According to the Frame Relay Forum, as many as 255 voice channels can be encoded over a single PVC, although the number is usually smaller when actually implemented.

Frame relay is a virtual circuit service. When a customer wants to connect two locations using frame relay, he or she contacts his or her service provider and tells the service representative the locations of the endpoints and the bandwidth he or she requires. The service provider issues a service order to create the circuit. If at some point in the future the customer decides to change the circuit endpoints or upgrade the bandwidth, another service order must be issued. This service is called PVC and is the most commonly deployed frame relay solution.

Frame relay is also capable of supporting *switched virtual circuit* (SVC) service, but SVCs are for the most part not available from service providers. With SVC service, customers can make their own modifications to the circuit by accessing the frame relay switch in the central office and requesting changes. However, service providers do not currently offer SVC service because of billing and tracking concerns (customer activities are difficult to monitor). Instead, they enable customers to create a fully meshed network between all locations for a very reasonable price. Instead of making routing changes in the switch, the customer has a circuit between every possible combination of desired endpoints. As a result, customers get the functionality of a switched network, while the service provider avoids the difficulty of administering a network within which the customer is actively making changes.

In frame relay, PVCs are identified using an address called a *Data Link Connection Identifier* (DLCI) (pronounced "del-sie"). At any given endpoint, the customer's router can support multiple DLCIs, and each DLCI can be assigned varying bandwidths based upon the requirements of the device/application on the router port associated with that DLCI. In Figure 1-60, the customer has purchased a T1 circuit to connect his or her router to the frame relay network. The router is connected to a videoconferencing unit at 384 Kbps, a frame-relay-capable PBX at 768 Kbps, and a data circuit for Internet access at 512 Kbps. Note that the aggregate bandwidth assigned to these devices exceeds the actual bandwidth of the access line by 128 Kbps (1,664 through 1,536). Under normal circumstances this would not be possible, but frame relay assumes that the traffic that it will normally be transporting is bursty by nature. If the assumption is correct (and it usually is), there is very little likelihood that all three devices will burst at the same instant in time. As a consequence, the circuit's operating capacity can actually be overbooked, a process known as *oversubscription*. Most service providers allow as much as 200 percent oversubscription, something customers clearly benefit from provided the circuit is designed properly. This means that the salesperson must carefully assess the nature of the traffic that the customer will be sending over the link and ensure that enough bandwidth is allocated to support the requirements of the various devices that will be sharing access to the link. Failure to do so can result in

Figure 1-60
Frame relay service delivery

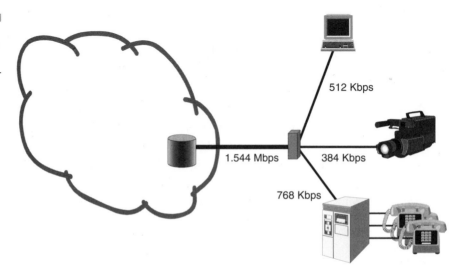

512 Kbps

1.544 Mbps   384 Kbps

768 Kbps

an underengineered facility that will not meet the customer's throughput requirements. This is a critical component of the service delivery formula.

The throughput level, that is, the bandwidth that frame relay service providers absolutely guarantee on a PVC-by-PVC basis, is called the *Committed Information Rate* (CIR). In addition to CIR, service providers will often support an *Excess Information Rate* (EIR), which is the rate above the CIR they will attempt to carry, assuming the capacity is available within the network. However, all frames above the CIR are marked as eligible for discard, which simply means that the network will do its best to deliver them but makes no guarantees. If push comes to shove and the network finds itself to be congested, the frames marked *discard eligible* (DE) are immediately discarded at their point of ingress. This CIR/EIR relationship is poorly understood by many customers because the CIR is taken to be an indicator of the absolute bandwidth of the circuit. Whereas bandwidth is typically measured in bits per second, CIR is a measure of bits in one second. In other words, the CIR is a measure of the average throughput that the network will guarantee. The actual transmission volume of a given CIR may be higher or lower than the CIR at any point in time because of the bursty nature of the data being sent, but in aggregate the network will maintain an average, guaranteed flow volume for each PVC. This is a selling point for frame relay. In most cases, customers get more than they actually pay for, and as long as the switch loading levels are properly engineered, the switch (and therefore the frame relay service offering) will not suffer adversely from this charitable bandwidth allocation philosophy. The key to success when selling frame relay is to have a very clear understanding of the applications the customer intends to use across the link so that the access facility can be properly sized for the anticipated traffic load.

Managing Service in Frame Relay Networks Frame relay does not offer a great deal of granularity when it comes to QoS. The only inherent mechanism is the DE bit described earlier as a way to control network congestion. However, the DE bit is binary. It has two possible values, which means that a customer has two choices: the information being sent is either important or it isn't. This is not particularly useful for establishing a variety of QoS levels. Consequently, a number of vendors have implemented proprietary solutions for QoS management. Within their routers (sometimes called *Frame Relay Access Devices* [FRADs]), they have established queuing mechanisms that enable customers to create multiple priority levels for differing traffic flows. For example, voice and video, which don't tolerate delay well, could be assigned to a higher-priority queue than the one to which asynchronous data traffic would be assigned. This enables frame relay to provide highly granular service. The downside is that this approach is proprietary, which means that the same vendor's equipment must be used on both ends of the circuit. Given the strong move toward interoperability, this is not an ideal solution because it locks the customer into a single vendor situation.

Congestion Control in Frame Relay   Frame relay has two congestion control mechanisms. Embedded in the header of each frame relay frame are two additional bits called the *Forward Explicit Congestion Notification* (FECN) bit and the *Backward Explicit Congestion Notification* (BECN) bit. Both are used to notify devices in the network of congestion situations that could affect throughput.

Consider the following scenario. A frame relay frame arrives at the second of three switches along the path to its intended destination, where it encounters severe local congestion (see Figure 1-61). The congested switch sets the FECN bit to indicate the presence of congestion and transmits the frame to the next switch in the chain. When the frame arrives, the receiving switch takes note of the FECN bit and tells it the following: "I just came from that switch back there, and it's extremely congested. You can transmit stuff back there if you want to, but there's a good chance that anything you send will be discarded, so you might want to wait awhile before transmitting." In other words, the switch has been notified of a congestion condition to which it may respond by throttling back its output to allow the affected switch time to recover.

On the other hand, the BECN bit is used to flow-control a device that is sending too much information into the network. Consider the situation shown in Figure 1-62, where a particular device on the network is transmitting at a high volume level, routinely violating the CIR and perhaps the EIR level established by mutual consent. The ingress switch—that is, the first switch the traffic touches—has the

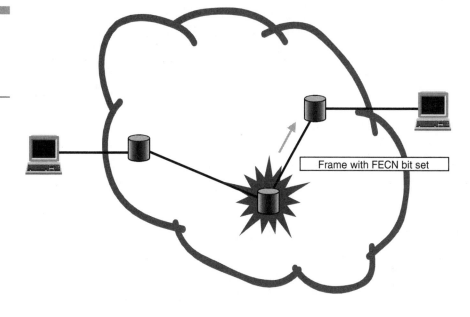

Figure 1-61
A frame
encounters
congestion in
transit

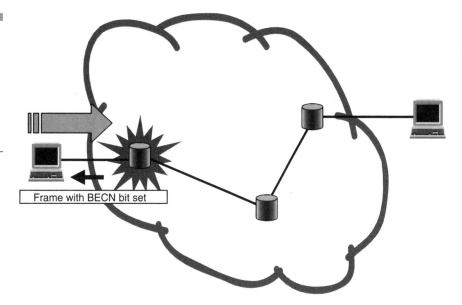

Figure 1-62
BECN bit used to
warn a station that
is transmitting too
much data (in
violation of its
agreement with
the switch)

capability to set the BECN bit on frames going toward the offending device, which carries the implicit message, "Cut it out or I'm going to hurt you." In effect, the BECN bit notifies the offending device that it is violating protocol and continuing to do so will result in every frame from that device being discarded—without warning or notification. If this happens, it gives the ingress switch the opportunity to recover. However, it doesn't fix the problem—it merely forestalls the inevitable because sooner or later the intended recipient will realize that frames are missing and will initiate recovery procedures, which will cause resends to occur. However, it may give the affected devices time to recover before the onslaught begins anew.

The problem with FECN and BECN lies in the fact that many devices choose not to implement them. They do not necessarily have the inherent capability to throttle back upon receipt of a congestion indicator, although devices that can are becoming more common. Nevertheless, proprietary solutions are in widespread use and will continue to be for some time to come.

Frame Relay Summary Frame relay is clearly a Cinderella technology, evolving quickly from a data-only transport scheme to a multi-service technology with diverse capabilities. For video, data, and some voice applications, it shines as a WAN offering. In some areas, however, frame relay is lacking. Its bandwidth is limited to DS3, and its capability to offer standards-based QoS is limited. Given the focus on QoS that is so much a part of customers' chanted mantra today and the flexibility that a switched solution permits, something else was required. That something was ATM.

Asynchronous Transfer Mode (ATM) Network architectures often develop in concert with the corporate structures that they serve. Companies with centralized management authorities such as utilities, banks, and hospitals often have centralized and tightly controlled hierarchical data-processing architectures to protect their data. On the other hand, organizations that are distributed in nature such as research and development facilities and universities often have highly distributed data-processing architectures. They tend to share information on a peer-to-peer basis and their corporate structures reflect this fact.

ATM came about not only because of the proliferation of diverse network architectures, but also because of the evolution of traffic characteristics and transport requirements. To the well-known demands of voice, we now add various flavors of data, video, MP3, an exponentially large variety of IP traffic, interactive real-time gaming, and a variety of other content types that place increasing demands on the network. Further, we have seen a need arise for a mechanism that can transparently and correctly transport the mix of various traffic types over a single network infrastructure, while at the same time delivering granular, controllable, and measurable QoS levels for each service type. In its original form, ATM was designed to do exactly that, working with SONET or SDH to deliver high-speed transport and switching throughout the network in the wide area, the metropolitan area, the campus environment, the LAN, right down to the desktop, seamlessly, accurately, and fast.

Today, because of competition from technologies such as QoS-aware IP transport, proprietary high-speed mesh networks, and Fast and Gigabit Ethernet, ATM has for the most part lost the race to the desktop. ATM is a cell-based technology, which simply means that the fundamental unit of transport—a frame of data, if you will—is of a fixed size, which enables switch designers to build faster, simpler devices because they can always count on their switched payload being the same size at all times. That cell comprises a 5-octet header and a 48-octet payload field, as shown in Figure 1-63. The payload contains user data; the header contains information that the network requires to both transport the payload correctly and ensure proper QoS levels for the payload. ATM accomplishes this task well, but at a cost. The 5-octet header comprises nearly 10 percent of the cell, a rather significant price to pay, particularly when other technologies such as IP and SONET add their own significant percentages of overhead to the overall payload. This reality is part of the problem. ATM's original claims to fame and the reasons it rocketed to the top of the technology hit parade were its capability to switch cells at tremendous speed through the fabric of the WAN and the ease with which the technology could be scaled to fit any network situation. Today, however, given the availability of high-speed IP

Figure 1-63
The ATM cell

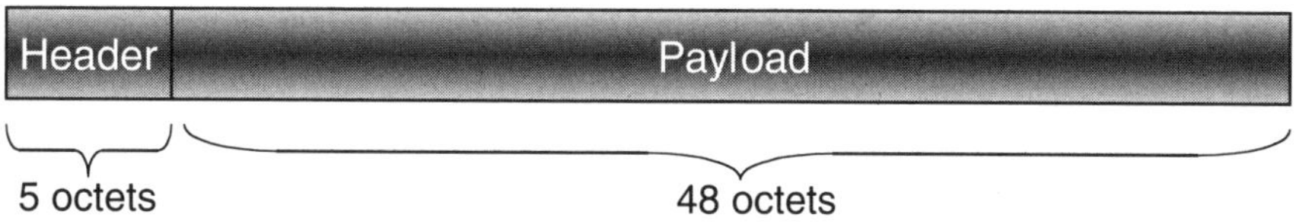

Header	Payload

5 octets   48 octets

routers that routinely route packets at terabit rates, ATM's advantages have begun to pale to a certain degree.

ATM Evolution   ATM has, however, emerged from the flames in other ways. Today, many service providers see ATM as an ideal aggregation technology for diverse traffic streams that need to be combined for transport across a WAN that will most likely be IP based. ATM devices, then, will be placed at the edge of the network, where they will collect traffic for transport across the Internet or (more likely) a privately owned IP network. Furthermore, because it has the capability to be something of a chameleon by delivering diverse services across a common network fabric, ATM is further guaranteed a seat at the technology game.

It is interesting to note that the traditional, legacy telecommunications network comprises two principle regions that can be clearly distinguished from each other: the network itself, which provides switching, signaling, and transport for traffic generated by customer applications, and the access loop, which provides the connectivity between the customer's applications and the network. In this model, the network is considered to be a relatively intelligent medium, whereas the customer equipment is usually considered to be relatively stupid.

Not only is the intelligence seen as being concentrated within the confines of the network, but the bulk of the bandwidth is also concentrated this way because the legacy model indicates that traditional customer applications don't require much of it. Between central office switches, however, and between the offices themselves, enormous bandwidth is required.

Today, this model is changing. Customer equipment has become remarkably intelligent, and many of the functions previously done within the network cloud are now performed at the edge. PBXs, computers, and other devices are now capable of making discriminatory decisions about required service levels, eliminating any need for the massive intelligence embedded in the core.

At the same time, the bandwidth is migrating from the core of the network toward the customer as applications evolve to require it. There is still massive bandwidth within the cloud, but the margins of the cloud are expanding toward the customer.

The result of this evolution is a redefinition of the network's regions. Instead of a low-speed, low-intelligence access area and a high-speed, highly intelligent core, the intelligence has migrated outward to the margins of the network, and the bandwidth, which was once exclusively a core resource, is now equally distributed at the edge. Thus, we see something of a core and edge distinction evolving as customer requirements change.

One reason for this steady migration is the well-known fact within sales and marketing circles that products sell best when they are located close to the buying customer. They are also easier to customize for individual customers when they are physically closest to the situation for which the customer is buying them.

ATM Technology Overview   Because ATM plays such a major role in networks today, it is important to develop at least a rudimentary understanding of its functions, architectures, and offered services.

ATM Protocols   Like all modern technologies, ATM has a well-developed protocol stack, shown in Figure 1-64, that clearly delineates the functional breakdown of the service. The stack consists of four layers: the Upper Services Layer, the ATM Adaptation Layer, the ATM Layer, and the Physical Layer.

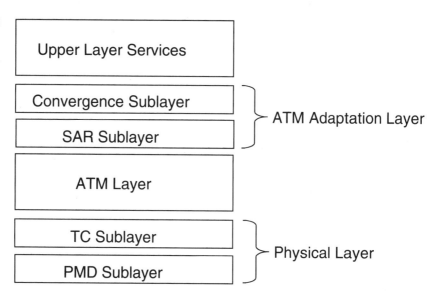

Figure 1-64
ATM protocol stack

The Upper Services Layer defines the nature of the actual services that ATM can provide. It identifies both *constant-bit-rate* (CBR) and *variable-bit-rate* (VBR) services. Voice is an example of a CBR service, while signaling, IP, and frame relay are examples of both connectionless and connection-oriented VBR services.

The *ATM Adaptation Layer* (AAL) has four general responsibilities:

- Synchronization and recovery from errors
- Error detection and correction
- Segmentation and reassembly of the data stream
- Multiplexing

The AAL comprises two functional sublayers. The *Convergence Sublayer* provides service-specific functions to the Services Layer so that the Services Layer can make the most efficient use of the underlying cell relay technology that ATM provides. Its functions include clock recovery for end-to-end timing management, a recovery mechanism for lost or out-of-order cells, and a timestamp capability for time-sensitive traffic such as voice and video.

The *Segmentation and Reassembly* (SAR) *Sublayer* converts the user's data from its original incoming form into the 48-octet payload chunks that will become cells. For example, if the user's data is arriving in the form of 64KB IP packets, SAR chops them into 48-octet payload pieces. It also has the responsibility to detect lost or out-of-order cells that the Convergence Sublayer will recover from and to detect single bit errors in the payload chunks.

The *ATM Layer* has five general responsibilities:

- Cell multiplexing and demultiplexing
- Virtual path and virtual channel switching
- Creation of the cell header
- Generic flow control
- Cell delineation

Because the ATM Layer creates the cell header, it is responsible for all of the functions that the header manages. The process, then, is fairly straightforward. The user's data passes from the Services

Layer to the ATM Adaptation Layer, which segments the data stream into 48-octet pieces. The pieces are handed to the ATM Layer, which creates the header and attaches it to the payload unit, thus creating a cell. The cells are then handed down to the Physical Layer.

The *Physical Layer* consists of two functional sublayers as well: the *Transmission Convergence* (TC) *Sublayer* and the *Physical Medium* (PMD) *Sublayer*. The TC Sublayer performs three primary functions. The first is called *cell rate decoupling*, which adapts the cell creation and transmission rate to the rate of the transmission facility by performing cell stuffing, which is similar to the bit stuffing process described earlier in the discussion of DS3 frame creation. The second responsibility is *cell delineation*, which enables the receiver to delineate between one cell and the next. Finally, it generates the transmission frame in which the cells are to be carried.

The PMD Sublayer takes care of issues that are specific to the medium being used for transmission, such as line codes, electrical and optical concerns, timing, and signaling.

The Physical Layer can use a wide variety of transport options, including

- DS1/DS2/DS3

- E1/E3

- 25.6 Mbps *user-to-network interface* (UNI) over UTP-3

- 51 Mbps UNI over UTP-5 (*Transparent Asynchronous Transmitter/Receiver Interface* [TAXI])

- 100 Mbps UNI over UTP-5

- OC3/12/48c

Others, of course, will follow as transport technologies advance.

The ATM Cell Header As we mentioned before, ATM is a cell-based technology that relies on a 48-octet payload field that contains actual user data and a 5-byte header that contains information needed by the network to route the cell and provide proper levels of service.

The ATM cell header, shown in Figure 1-65, is examined and updated by each switch it passes through and comprises six distinct fields: the *Generic Flow Control* (GFC) field, the *Virtual Path*

Figure 1-65
ATM cell header
details

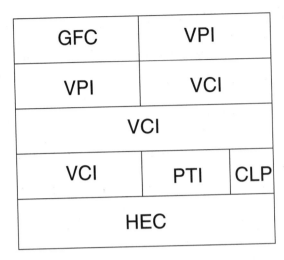

GFC	VPI	
VPI	VCI	
VCI		
VCI	PTI	CLP
HEC		

Identifier (VPI) field, the *Virtual Channel Identifier* (VCI) field, the *Payload Type Identifier* (PTI) field, the *Cell Loss Priority* (CLP) field, and the *Header Error Control* (HEC) field:

- *GFC* This 4-bit field is used across the UNI for network-to-user flow control. It has not yet been completely defined in the ATM standards, but some companies have chosen to use it for very specific purposes. For example, Australia's Telstra Corporation uses it for flow control in the network-to-user direction and as a traffic priority indicator in the user-to-network direction.

- *VPI* The 8-bit VPI identifies the virtual path over which the cells will be routed at the UNI. It should be noted that because of dedicated, internal flow control capabilities within the network, the GFC field is not needed across the *network-to-network interface* (NNI). It is therefore redeployed. The four bits are converted to additional VPI bits, thus extending the size of the virtual path field. This allows for the identification of more than 4,000 unique virtual paths. At the UNI, this number is excessive, but across the NNI, it is necessary because of the number of potential paths that might exist between the switches that make up the fabric of the network.

- *VCI* As the name implies, the 16-bit VCI identifies the unidirectional virtual channel over which the current cells will be routed.

- *PTI* The 3-bit PTI field is used to indicate network congestion and the cell type in addition to a number of other functions. The first bit indicates whether the cell was generated by the user or by the network, while the second indicates the presence or absence of congestion in user-generated cells or flow-related *Operations, Administration, and Maintenance* (OA&M) information in cells generated by the network. The third bit is used for service-specific, higher-layer functions in the user-to-network direction, for example, to indicate that a cell is the last in a *series* of cells. From the network to the user, the third bit is used with the second bit to indicate whether the OA&M information refers to segment or end-to-end-related information flow.

- *CLP* The single-bit CLP field is a relatively primitive flow control mechanism by which the user can indicate to the network which cells to discard in the event of a condition that demands that some cells be eliminated, similar to the DE bit in frame relay. It can also be set by the network to indicate to downstream switches that certain cells in the stream are eligible for discard should that become necessary.

- *HEC* The 8-bit HEC field can be used for two purposes. First, it provides for the calculation of an 8-bit CRC that checks the integrity of the entire header. Second, it can be used for cell delineation.

Addressing in ATM ATM is a connection-oriented, virtual circuit technology, meaning that communication paths are created through the network prior to actually sending traffic. Once established, the ATM cells are routed based upon a virtual circuit address. A virtual circuit is simply a connection that gives the user the appearance of being dedicated to that user, when in fact the only thing that is actually dedicated is a time slot. This technique is generically known as *label-based switching* and is accomplished through the use of routing tables in the ATM switches that designate input ports, output ports, input addresses, output addresses, and QoS parameters required for proper routing and service provisioning. As a result, cells do not contain explicit destination addresses, but rather contain time slot identifiers.

There are two components to every virtual circuit address, as shown in Figure 1-66. The first is the *virtual channel*, which is a unidirectional conduit for the transmission of cells between two endpoints. For example, if two parties are conducting a videoconference, they will each have a virtual channel for the transmission of outgoing cells that make up the video portion of the conference.

The second level of the ATM addressing scheme is called a *virtual path*. A virtual path is a bundle of virtual channels that have the same endpoints and that when considered together make up a bidirectional transport facility. The combination of unidirectional channels that we need in our two-way videoconferencing example makes up a virtual path.

ATM Services   The basic services that ATM provides are based on three general characteristics: the nature of the connection between the communicating stations (connection-oriented versus connectionless), the timing relationship between the sender and the receiver, and the bit rate required to ensure proper levels of service quality. Based on those generic requirements, both the ITU-T and the ATM Forum have created service classes that address the varying requirements of the most common forms of transmitted data.

ITU-T Service Classes   The ITU-T assigns service classes based on three characteristics: connection mode, bit rate, and the end-to-end timing relationship between the end stations. They have created four distinct service classes based on the model shown in Figure

Figure 1-66
Addressing in ATM with virtual channels and paths

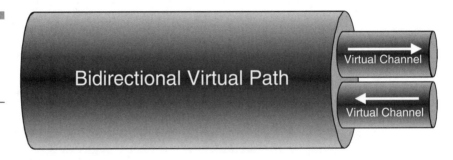

Bidirectional Virtual Path

Virtual Channel

Virtual Channel

1-67. Class A service, for example, defines a connection-oriented, CBR, timing-based service that is ideal for the stringent requirements of voice service. Class B, on the other hand, is ideal for services such as VBR video in that it defines a connection-oriented, VBR, timing-based service.

Class C service is defined for things such as frame relay in that it provides a connection-oriented, VBR, timing-independent service. Finally, Class D delivers a connectionless, VBR, timing-independent service that is ideal for IP traffic as well as the *Switched Multimegabit Data Service* (SMDS).

In addition to service classes, the ITU-T has AAL service types, which align closely with the A, B, C, and D service types described previously. Whereas the service classes (A, B, C, and D) describe the capabilities of the underlying network, the AAL types describe the cell format. They are AAL1, AAL2, AAL3/4, and AAL 5. However, only two of them have really survived in a big way.

AAL1 is defined for Class A service, which is a CBR environment ideally suited for video and voice applications. In AAL1 cells, the first octet of the payload serves as a payload header that contains cell sequence and synchronization information that is required to provision CBR, fully sequenced service. AAL1 provides *circuit emulation service* (CES) without dedicating a physical circuit, which explains the need for an end-to-end timing relationship between the transmitter and the receiver.

Figure 1-67

ITU-T ATM service definitions

	Class A	Class B	Class C	Class D
AAL Type	1	2	5, 3/4	5, 3/4
Connection Mode	Connection-oriented	Connection-oriented	Connection-oriented	Connection-less
Bit Rate	Constant	Variable	Variable	Variable
Timing Relationship	Required	Required	Not required	Not required
Service Types	Voice, video	VBR voice, video	Frame relay	IP

AAL5, on the other hand, is designed to provide both Class C and D services, and although it was originally proposed as a transport scheme for connection-oriented data services, it turns out to be more efficient than AAL3/4 and accommodates connectionless services quite well.

To guard against the possibility of errors, AAL5 has an 8-octet trailer appended to the user data that includes a variable-size *pad field* used to align the payload on 48-octet boundaries, a 2-octet *control field* that is currently unused, a 2-octet *length field* that indicates the number of octets in the user data, and finally a 4-octet *CRC* that can check the integrity of the entire payload. AAL5 is often referred to as the *Simple and Easy Adaptation Layer* (SEAL), and it may find an ideal application for itself in the burgeoning Internet arena. Recent studies indicate that TCP/IP transmissions produce comparatively large numbers of small packets that tend to be around 48 octets long. That being the case, AAL5 could transport the bulk of them in its user data field. Furthermore, the maximum size of the user data field is 65,536 octets, which is coincidentally the same size as an IP packet.

ATM Forum Service Classes   The ATM Forum looks at service definitions slightly differently than the ITU-T, as shown in Figure 1-68. Instead of the A, B, C, and D services, the ATM Forum categorizes them as real-time and nonreal-time services. Under the real-

Figure 1-68
ATM Forum service definitions

Service	Descriptors	Loss	Delay	Bandwidth	Feedback
CBR	PCR, CDVT	Yes	Yes	Yes	No
VBR-RT	PCR, CDVT, SCR, MBS	Yes	Yes	Yes	No
VBR-NRT	PCR, CDVT, SCR, MBS	Yes	Yes	Yes	No
UBR	PCR, CDVT	No	No	No	No
ABR	PCR, CDVT, MCR	Yes	No	Yes	Yes

time category, they define CBR services that demand fixed resources with guaranteed availability. They also define real-time VBR service, which provides for statistical multiplexed, variable-bandwidth service allocated on demand. A further subset of real-time VBR is peak-allocated VBR, which guarantees constant loss and delay characteristics for all cells in that flow.

Under the nonreal-time service class, *unspecified bit rate* (UBR) is the first service category. UBR is often compared to IP in that it is a best-effort delivery scheme in which the network provides whatever bandwidth it has available, with no guarantees made. All recovery functions from lost cells are the responsibility of the end-user devices.

UBR has two subcategories of its own. The first, *nonreal-time VBR* (NRT-VBR), improves the impacts of cell loss and delay by adding a network resource reservation capability. *Available bit rate* (ABR), UBR's other subcategory, makes use of feedback information from the far end to manage loss and ensure fair access to and transport across the network.

Each of the five classes makes certain guarantees with regard to cell loss, cell delay, and available bandwidth. Furthermore, each of them takes into account descriptors that are characteristic of each service described. These include *peak cell rate* (PCR), *sustained cell rate* (SCR), *minimum cell rate* (MCR), *cell delay variation tolerance* (CDVT), and *burst tolerance* (BT).

ATM Forum Specified Services   The ATM Forum has identified a collection of services for which ATM is a suitable, perhaps even desirable, network technology. These include *cell relay service* (CRS), CES, *voice and telephony over ATM* (VTOA), *frame relay bearer service* (FRBS), *LAN emulation* (LANE), MPOA, and VOD.

CRS is the most basic of the ATM services. It delivers precisely what its name implies: a "raw pipe" transport mechanism for cell-based data. As such, it does not provide any ATM bells and whistles, such as QoS discrimination; nevertheless, it is the most commonly implemented ATM offering because of its lack of implementation complexity.

CES gives service providers the ability to offer a selection of bandwidth levels by varying both the number of cells transmitted per second and the number of bytes contained in each cell.

VTOA is a service that has yet to be clearly defined. The ability to transport voice calls across an ATM network is a nonissue, given the availability of Class A service. What is not clearly defined, however, are corollary services such as 800/888 calls, 900 service, 911 call handling, enhanced services billing, SS7 signal interconnection, and so on. Until these issues are clearly resolved, ATM-based, feature-rich telephony will not become a mainstream service, but will instead be limited to simple voice—and there is a difference.

FRBS refers to the capability of ATM to interwork with frame relay. Conceptually, the service implies that an interface standard enables an ATM switch to exchange data with a frame relay switch, thus allowing for interoperability between frame and cell-based services. However, many manufacturers are taking a slightly different view. They build switches with soft, chewy cell technology at the core and surround the core with hard, crunchy interface cards to suit the needs of the customer.

For example, an ATM switch might have ATM cards on one side to interface with other ATM devices in the network, and frame relay cards on the other side to enable it to communicate with other frame relay switches, as shown in Figure 1-69. Thus, a single piece of hardware can logically serve as both a cell and frame relay switch. This

Figure 1-69
FRBS in ATM

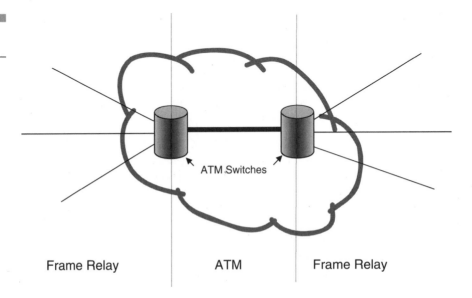

Frame Relay ATM Frame Relay

design is becoming more and more common because it helps to avoid a future rich with forklift upgrades.

LANE enables an ATM network to move traffic transparently between two similar LANs, but it also enables ATM to transparently slip into the LAN arena. For example, two Ethernet LANs could communicate across the fabric of an ATM network, as could two Token Ring LANs. In effect, LANE enables ATM to provide a bridging function between similar LAN environments. In LANE implementations, the ATM network does not handle MAC functions such as collision detection, token passing, or beaconing; it merely provides the connectivity between the two communicating endpoints. The MAC frames are transported inside AAL5 cells.

One clear concern about LANE is that LANs are connectionless, whereas ATM is a virtual circuit-based, connection-oriented technology. LANs routinely broadcast messages to all stations, while ATM allows point-to-point or multipoint circuits only. Thus, ATM must look like a LAN if it is to behave like one. To make this happen, LANE uses a collection of specialized LAN emulation clients and servers to provide the connectionless behavior expected from the ATM network.

On the other hand, MPOA provides the ATM equivalent of routing in LAN environments. In MPOA installations, routers are referred to as MPOA servers. When one station wants to transmit to another station, it queries its local MPOA server for the remote station's ATM address. The local server then queries its neighbor devices for information about the remote station's location. When a server finally responds, the originating station uses the information to establish a connection with the remote station, while the other servers cache the information for further use.

MPOA promises a great deal, but it is complex to implement and requires other ATM components such as the private NNI capability to work properly. Furthermore, it's being challenged by at least one alternative technology—*IP switching*.

IP switching is far less overhead intensive than MPOA. Furthermore, it takes advantage of a known (but often ignored) reality in the LAN interconnection world: most routers today use IP as their core protocol, and the great majority of LANs are still Ethernet. This means that a great deal of simplification can be done by crafting networks to operate around these two technological bases. In fact, this is precisely what IP switching does. By using existing, low-overhead protocols, the IP switching software creates new ATM connections dynamically and quickly, updating switch tables on the fly. In IP switching environments, IP resides on top of ATM within a device, as shown in Figure 1-70, providing the best of both protocols. If two communicating devices want to exchange information and they have done so before, if ATM mapping already exists, and if layer 3 involvement (IP) is required, the ATM switch portion of the service simply creates the connection at high speed. If an address lookup is required, then the call is handed up to IP, which takes whatever steps are required to perform the lookup (a DNS request, for example). Once it has the information, it hands it down to ATM, which proceeds to set up the call. The next time the two need to communicate, ATM will be able to handle the connection.

There are other services looming on the horizon in which ATM plays a key role, one of which is VOD. This leads to what I often refer to as "the great triumvirate:" ATM, SONET or SDH, and broadband services. By combining the powerful switching and multiplexing fab-

Figure 1-70
IP switching

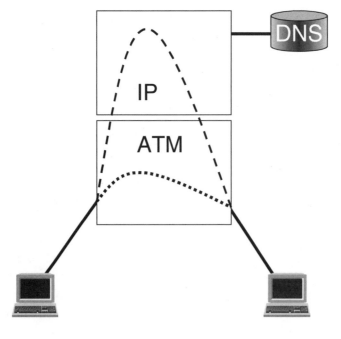

ric of ATM with the limitless transport capabilities of SONET or SDH, true broadband services can be achieved, and the idea of creating a network that can be all things to all services can finally be realized.

The Internet Protocol (IP)   IP is part of the well-known TCP/IP protocol suite that came about in the 1960s as an integral part of the Department of Defense's ARPANET project. It is a network layer protocol that is responsible for routing functions and congestion control and works closely with its transport layer counterpart, TCP. TCP/IP supports a wide array of application services and will operate over many different physical media types.

IP is a connectionless network layer protocol, which means that it does not perform a call setup prior to transmitting data packets. Instead, it transmits the packets into the network, each bearing a complete destination address, and trusts the network to deliver them to their final destination. With the help of various routing protocols, the packets do generally arrive, but because the service is connectionless and because connectionless networks do not discern a relationship between packets that originate from the same source, they may take different routes from source to destination based on changing network congestion and other factors. Consequently, they may arrive out of sequence. The advantage of a connectionless network, of course, is that it has the capability to avoid troubled areas in the network by dynamically routing around them. This flexibility is worth the cost of connectionless service, particularly since the transport layer (TCP) will generally correct any discrepancies that arise from IP's connectionless nature.

Addressing in IP   The current version of IP, known as *IP Version 4* (IPv4), relies on what is known as a *dotted decimal addressing scheme*. The name derives from the fact that IP addresses comprise 4 8-bit segments separated by periods, as shown in the following: 255.255.255.255.

Each IP address is 32 bits long (4 8-bit components), which means that 2^{32} (roughly 4.2 billion) possible addresses can be created. The good news is that 4.2 billion is a lot. The bad news is that it isn't enough! Because of the immense and unexpected popularity of IP as a universal addressing scheme, we are getting dangerously close to exhausting the available address space. There are a number of reasons for this: the diversity of devices that can actually be addressed by IP is growing, networked devices in general are growing, and the manner in which IP addresses are assigned is not particularly efficient.

To understand the challenges of IP addressing, it is first necessary to understand how the addresses work. There are five classes of IP addresses, labeled A, B, C, D, and E. Classes A through C are used commercially, while Class D is reserved for multicasting and Class E is reserved for future use. They are created as follows. In any given company, it is necessary to identify both the network that serves that company as well as the devices attached to that network. By assigning the four 8-bit pieces of the IP address in a variety of ways, we can create a tiered address scheme that enables us to identify a small number of addresses with many hosts (devices), a large number of networks with a few hosts, or anything in between. Consider the illustrations in Figure 1-71. In a Class A address, three of the bytes are assigned as host identifiers, while one is used to identify networks. Consequently, a Class A address can identify as many as 16 million unique hosts (2^{24}) and up to 126 ($2^7 - 2$) unique networks. As it turns out, not all of the bits in each byte are actually used for addressing—some are reserved to indicate the class of the address. For example, in Class A, the most significant bit is set to 0 to indicate

Figure 1-71
IP addressing
scheme

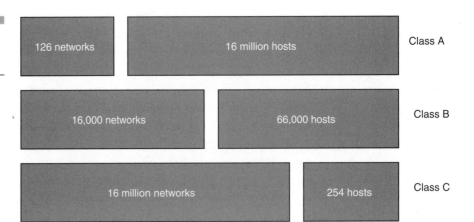

that it is, in fact, a Class A address, leaving 7 bits to identify the network. Some addresses are reserved so the maximum number of identifiable networks in a Class A address is 26. Similarly, a Class B address can identify 65,534 hosts on 16,382 networks, while a Class C address, the most commonly deployed of all, can identify 254 hosts on each of more than 2 million networks.

The biggest problem facing IP today is the fact that when IP addresses are assigned to corporations, they are generally assigned an entire address block based on the number of employees/devices they must provide with a static network address. For example, a company of 50 people might acquire an entire Class C address, which would enable them to easily accommodate their 50 employees, but would leave a significant number of IP addresses within the range idle. This is a serious problem. Some reports estimate that as few as 20 percent of all assigned IP addresses are actually in use.

To combat this problem, a number of solutions have been devised. The first we will describe is called *subnet masking*. In subnet masking, IP addresses are broken into three pieces instead of two (NETID and HOSTID). In subnet masking, the first piece identifies the network, the second identifies a subnet address, and the third identifies the host on that subnet. Each subnet is limited to 254 nodes, but within a Class B address, for example, there can be as many as 256 uniquely identifiable subnets. In effect, the network is subdivided into pieces, which makes management and administration easier—and far more granular. By creating a larger number of unique networks within the address space normally reserved for one network, a better assignment of IP addresses is possible.

The second technique that is being used is a protocol called the *Dynamic Host Configuration Protocol* (DHCP). DHCP enables a server to dynamically assign IP addresses to requesting stations, thus eliminating the problem of unused IP address space.

The third solution is called *Classless InterDomain Routing* (CIDR). CIDR (pronounced "ci-der") enables Class C addresses to be broken into smaller pieces than 254 hosts. This helps to resolve the problem of IP address overassignment. If a company buys a Class C block, for example, and later determines that it needs another 50 addresses, it must actually acquire another entire Class C address. With CIDR, they can buy small pieces.

Finally, the solution that many are waiting for is the next version of IP, called *IP Version 6* (IPv6).[4] Sometimes called IPng (for *next generation*), IPv6 adds enhancements to IPv4, including 128-bit addressing, dynamic reconfiguration, enhanced network security, multicasting, better QoS discrimination, and a number of other features. With 128 bits of addressing space, 2^{128} unique addresses can be created, which is approximately 10^{38} in total. One difficulty that has been identified is the conversion from IPv4 to IPv6. Many experts believe that the conversion is a necessary evil that must be faced and that the deployment of *Network Address Translation* (NAT) devices, which will convert between IPv4 and IPv6 addresses, will be required for some time. Indications from the so-called early adopters and implementers of IPv6 are that the advantages far outweigh the downsides of the conversion.

The Virtual Private Network (VPN)   One technology that facilitates the deployment of IP videoconferencing is the VPN, a recently adopted set of technologies that enables a network user to eliminate the cost of leased facilities without sacrificing the safety and security that they provide. VPNs work as follows. Instead of leasing or owning dedicated facilities between enterprise locations as a way to ensure the privacy of communications, service providers offer VPN service as a secure alternative. Rather than dedicated facilities, user traffic is carried across an IP-based public or corporate network with traffic from many other users. Each flow of data, however, is isolated from all others through the use of secure protocols that prevent one user from accessing the traffic of another, thus ensuring privacy and security.

This network model has several advantages for both the customer and service provider. The customer is granted a secure network that guarantees the privacy of transmitted data, yet is not saddled with the cost or complexity of a dedicated network infrastructure. The service provider has the remarkable advantage of being able to resell the same physical network over and over again to multiple different customers, thus giving them the ability to garner enhanced revenues

---

[4]IPv6 is described in IETF RFC 1752.

from a highly capital intensive resource. So everybody wins. In October 2001, the service provider Global Crossing announced a suite of global VPN services called SmartRoute and ExpressRoute. SmartRoute offers secure IP encryption, class of service management, and bandwidth management. It is considered a network-based VPN because all of the intelligence lies within the Global Crossing network. The service offers support for frame relay, private line, and ATM, and provides a seamless migration path for customers who want to evolve to an IP infrastructure.

ExpressRoute, on the other hand, relies on MPLS to create dedicated paths through the network. Both services are targeted at customers with high-bandwidth requirements.

For corporations that want to migrate to an IP infrastructure, the VPN is a good migratory path because it enables videoconferencing traffic to be carried across a shared network while still preserving the integrity of all components of the traffic mix. Because the video stream travels within its own logically segregated virtual channel, bandwidth contention does not occur.

Terminating the Signal

Compression schemes do a good job of compressing and faithfully reconstituting images, particularly when the image being compressed is a photograph or video clip. To understand the dynamics of this relationship, let's take a moment to consider what it takes to create a digital photograph displayed on a computer screen.

A typical laptop computer display is often referred to as being 640 × 480, 800 × 600, or 1,024 × 768. These numbers refer to the number of picture elements (more commonly called *pixels*) that make up the display, as illustrated in Figure 1-72. Look closely at the screen and you will find that it is made up of thousands of tiny spots of light (the pixels), each of which can take on whatever characteristics are required to correctly and faithfully paint the image on the screen. These characteristics include color components (sometimes called *chrominance*), black-and-white components (sometimes called *luminance*), brightness (the intensity of the signal), and hue (the actual

Figure 1-72
Pixels on a typical
computer screen

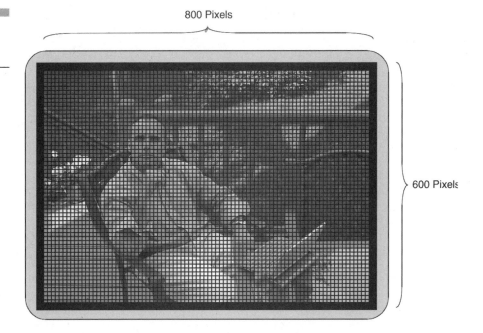

800 Pixels

600 Pixels

wavelength of the color). These characteristics are important in video and digital imaging systems because they determine the quality of the final image. The image, then, is a mosaic of light; the tiles that comprise the mosaic are *light-emitting diodes* (LEDs) that create the proper light at each pixel location.

Each pixel has a red, green, and blue (RGB) *light generator*, as shown in Figure 1-73. The colors red, green, and blue are called the *primary colors* because they form the basis for the creation of all other colors. It is a well-known fact that if three white lights are covered with red, green, and blue color gels, respectively, and the lights are shined at roughly the same spot, as shown in Figure 1-74, the result will be a light spot for each color, but the intersection of the three colors will be white light. The combination of the three primary colors creates white.

Each primary color also has a *complimentary color* in the overall spectrum. As Figure 1-75 shows, the complementary color for red is cyan, and the complimentary colors for green and blue are magenta and yellow, respectively. Table 1-3 shows the relationships that exist between the primary and complementary colors.

Figure 1-73
The RGB light emitters in a pixel

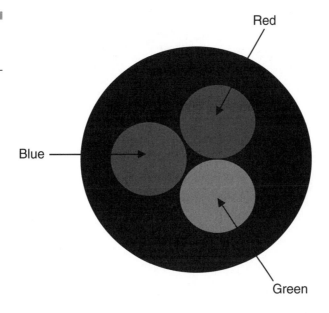

Red

Blue

Green

Figure 1-74
Red, green, and blue light, when combined, generate white light.

Table 1-3

Primary and complementary color combinations

If You Combine	The Result Is
Red + blue	Magenta
Green + red	Yellow
Blue + green	Cyan
Red + green + blue	White
Any two complementary colors	White

Figure 1-75
Complementary
colors

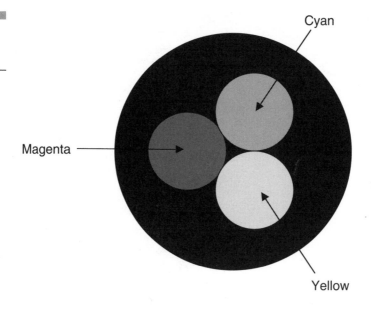

For full-color, uncompressed images, each of the RGB elements requires 8 bits, or 24 bits per pixel. This yields what is known as *256-bit color* (2^8). Now consider the storage requirements for an image that is 640 × 480 pixels in size:

$$640 \times 480 = 307{,}200 \text{ pixels}$$

$$307{,}200 \text{ pixels} \times 24 \text{ bits/pixel} = 7{,}372{,}800 \text{ bits}$$

$$7{,}372{,}800 \text{ bits}/8 \text{ bits per byte} = 921{,}600 \text{ bytes per image}$$

In other words, an uncompressed, relatively low-quality image requires 1MB of storage. A larger 1,024 × 768 pixel image requires 6.3MB of storage capacity in its uncompressed form. Given today's relatively low-bandwidth access solutions (such as ISDN, DSL, cable modems, and wireless), we should be thankful that compression technologies exist to reduce the bandwidth required to move them through a network!

Image Coding Schemes

Images are encoded using a variety of techniques. These include Windows *Bitmap* (BMP), the *Joint Photographic Experts Group* (JPEG), the *Graphical Interchange Format* (GIF), and the *Tag Image File Format* (TIFF). They are described in detail in the following sections.

Windows Bitmap (BMP) Windows BMP files are stored in a device-independent BMP format that enables the Windows operating system to display the image on any type of display device. The phrase "device independent" means that the bitmap specifies pixel color in a format that is independent of the method used by a display device to represent color. The default filename extension of a Windows file is .BMP.

The Joint Photographic Experts Group (JPEG) JPEG is a standard designed to control image compression. The term is named after the original name of the international body that created the standard. Comprised of both technologists and artists, the JPEG committee created a highly complete and flexible standard that compresses both full-color and grayscale images. It is effective when used with photographs, artwork, and medical images, and works less effectively on text and line drawings. JPEG is designed for the compression of still images, although there is a related standard (still in draft at the time of this writing) called *Motion JPEG 2000* (MJP2). MJP2 is not really a formal standard, although various vendors and developers have tried to formalize it. MPEG is designed for the compression of moving images such as multimedia, movies, and real-time diagnostic images. However, according to the MJP2 committee, MJP2 will be the compression technology of choice for medical imaging, security systems, and digital cameras.

JPEG is considered to be a *lossy* solution, meaning that once the original image has been compressed, the decompressed image loses a slight degree of integrity when it is viewed. There are, of course, lossless compression algorithms.

JPEG, however, achieves more efficient compression than is possible with competing lossless techniques. At first glance, this would appear to be a problem for many applications since the compression process actually squeezes information out of the original image, leaving a slightly inferior artifact. For example, a diagnostician examining a medical image might be concerned with the fact that the compressed image is not as good as the original. Luckily, this is not a problem for one very simple reason: the human eye is an imperfect viewing device. JPEG is designed to take advantage of well-understood limitations in the human eye, particularly the fact that small color changes are not perceived as discretely as small brightness changes. Clearly, JPEG is designed to compress images that will be viewed by people and therefore do not have to be absolutely faithful reproductions of the original image from which they were created. A machine analysis of a JPEG image would certainly yield inferior results, but it is perfectly acceptable to the human eye.

Of course, the degree of loss that occurs during a JPEG transformation can be controlled by adjusting a variety of compression parameters relative to each other. For example, file size and image quality can be adjusted relative to one another. A medical image, which requires extremely high quality on the image output side, would require a large file size, whereas a compressed text document could easily suffer significant loss without losing readability, resulting in a very small file.

The hardware or software coders that create (compress) and expand (decompress) JPEG images are called CODECs. If high image quality is not critically important, a low-cost, higher-loss CODEC can be used, thus reducing the overall cost of the deployed hardware or software solution.

JPEG is used as a compression tool for two primary reasons: to reduce file sizes for transmission or archival storage and to archive 24-bit color images instead of 8-bit color images. Clearly, the ability to reduce the number of bytes required to store an image is an advantage. It reduces the cost of networking by reducing the amount of connect time required to transmit an image across a network and lowers IT costs by reducing the amount of disk space required to store the image. JPEG can easily achieve compression ratios in excess of 20:1, meaning that a 2MB file becomes a 100KB entity following JPEG compression.

The second fundamental advantage of JPEG is that it stores full, 24-bit color information. GIF, the other image format that is widely used on the Web (discussed in the following section), stores 8 bits per bits/pixel—256-bit color. GIF is designed for inexpensive computer displays. However, high-end display units are becoming quite cost-effective, and JPEG photos are richer than GIF images when displayed on lower-cost displays. For this reason, GIF is seen by many as becoming obsolete.

The truth is, however, that JPEG will not completely replace GIF because GIF continues to be a superior solution for certain forms of images. For the most part, JPEG is better than GIF for storing full-color or grayscale images of natural scenes such as scanned photographs and continuous-tone artwork. Any smooth variation in color, such as what occurs in the highlighted or shaded areas of a subtle image, will be represented far more faithfully and in less space by JPEG.

On the other hand, GIF does a better job on images with only a few distinct colors such as sketches, maps, line drawings, and cartoons. Not only is GIF considered lossless for these images, it often achieves compression ratios that are much higher than JPEG can achieve. For example, large numbers of clustered pixels that are the same color are compressed very efficiently by GIF. JPEG, however, has a hard time compressing such data without introducing visible defects.

Computer images, such as vector drawings or ray-traced graphics, are typically found somewhere between photographs and line drawings in terms of their complexity in the eyes of the compression algorithm. The more complex the image, the more likely JPEG will be able to achieve significant levels of compression of the image. This is equally true with natural artwork. On the other hand, icons made up of only a few colors are better handled by GIF.

JPEG has difficulty achieving satisfactory compression with long, sharp edges. For example, if the image to be compressed has a row of black pixels immediately adjacent to a row of white pixels, the edges will often appear blurred unless a very high-quality setting is used, which, of course, reduces the degree of compression that JPEG will attempt to achieve. The good news is that such long, sharp edges are relatively uncommon in scanned photographs, but are common in GIF files. Straight lines are rare in nature for the subject of most photographs. They are quite common, however, in line drawings and illustrations. As a result, GIF is typically a better choice for the compression of these image types.

As a general rule, two-level black-and-white images should not be converted to JPEG because they (by nature) violate the conditions listed previously. Grayscale images that have 16 gray levels are far more acceptable to JPEG. However, GIF is considered to be a lossless encoding scheme for images of up to 256 levels, while JPEG continues to be lossy.

How JPEG Works The actual mechanical and mathematical processes that govern JPEG's inner workings are quite complex and will not be covered in detail here. Readers interested in more detail are directed to read the tutorial information that can be found at http://www.dcs.ed.ac.uk/home/mxr/gfx/2d/JPEG.txt. However, a high-level description of the process follows.

Earlier, we discussed the fact that a 640 × 480 pixel image requires nearly a full megabyte of disk space for uncompressed storage. To reduce that requirement, JPEG performs a mathematical permutation of the image that removes redundant information contained in the image, thus reducing storage requirements for the image. This results in loss of information, and although this appears to be a bad thing, it really isn't because of the limitations of the human eye described earlier.

Consider the image of my good friend Dennis McCooey shown in Figure 1-76. Using a high-quality digital still camera, I took this portrait of him shown in the illustration. I then enlarged a small section of the image to illustrate how the JPEG compression algorithm actually works. Keep in mind that a digital image comprises a large number of individual pixels, each of which has individual color characteristics associated with it. In the original photograph, Dennis' skin tones are relatively uniform across his face except for the areas that are in shadow, such as his eyes and parts of his chin. However, when we enlarge a small area of the image as shown in the lower right, the pixels vary widely in terms of their color components. This disparity is what JPEG exploits as a way of getting the job done. It knows that the human eye cannot see the difference between these pixels until they are dramatically enlarged as I have done in Figure 1-77.

Figure 1-76

Pixel group in a typical image. JPEG views each group as a separate entity.

Pixel group ←

What JPEG does is cluster the pixels in 16-by-16 pixel groups. It then mathematically removes every other pixel, resulting in 8-by-8 pixel arrays. It then performs a mathematical transformation of the 64-pixel arrays called a *discrete cosine transform* that calculates an average value for each group of pixels. Additional steps in the process further reduce the pixel count, resulting in a dramatic reduction in the total number of pixels required to represent the original image—hence, the term *lossy*. To restore the image, the reverse process is invoked.

JPEG is a 24-bit color scheme that yields high degrees of color fidelity. Unfortunately, many users do not have access to 24-bit color monitors, which means that a JPEG-encoded image will not display the image correctly. To display a full-color image, the host computer must analyze the file, choose an appropriate set of representative colors from a color chart, and map the image into the new color domain. This process is known as *color quantization.*

The speed and image quality of a JPEG viewer used on a lower-quality machine is largely determined by its quantization algorithm. If the algorithm is complex and well designed, the user will see

Figure 1-77
Pixels in a JPEG
image

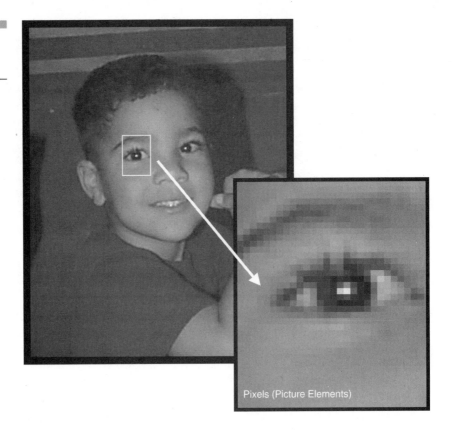

Pixels (Picture Elements)

minimal impact on the image from the quantization process. If the encoder is inferior, however, the result will be equally inferior.

Of course, there are ways around this. GIF images are already quantized to 256 or fewer colors. (GIF doesn't allow more than 256 color palette entries.) GIF has the advantage that the creator of the image calculates the degree of color quantization so that viewers do not have to. For this reason, GIF viewers are typically faster than JPEG viewers. Of course, this also limits the user to whatever constraints the GIF viewer places on the image. If the creator of the image quantized the original image to a level that cannot be achieved by a receiving device, the image will have to be further quantized, resulting in poorer image quality.

GIF clearly offers some advantages over JPEG, but in the long term JPEG offers better image quality than GIF. Of course, 8-bit displays are rapidly disappearing from the landscape in favor of higher-resolution monitors. For them, GIF has already become an academic argument because JPEG is a far better solution for image display.

An Aside: Progressive JPEG   Traditional JPEG compression schemes scan the image from top to bottom in a single effort. In Progressive JPEG, the file is divided into a series of scans. The first scan creates a low-quality image and requires minimal storage capacity. Each additional scan incrementally improves the quality of the original image. The advantage of progressive JPEG is that if an image must be viewed in real time as it is being transmitted, it is possible to see a low-quality rendition of the image very quickly and increasingly better images as further scans are performed. The disadvantage, of course, is that each scan requires the same amount of time to compute and display as a standard baseline JPEG.

Until quite recently, few applications lent themselves to Progressive JPEG. With the growing use of the Web for image transport, however, and the increasing availability of access bandwidth and high-speed PCs, Progressive JPEG has become quite popular.

Another Aside: Lossless JPEG   The JPEG standard defines a variety of compression techniques, one of which is called *lossless JPEG*. In fact, a new lossless standard is now available. JPEG-LS is a new lossless or near-lossless compression standard for continuous-tone images that is defined by the ISO-14495-1/ITU-T.87 standard. It is based on the *Low Complexity Lossless Compression for Images* (LOCO-I) algorithm developed at Hewlett-Packard Laboratories. It is described in more detail in the following section.

Lossless actually means "mathematically lossless," a technique that defines an algorithm that guarantees that its decompressed output file is identical to the original input. This seems to be a bit of a misnomer. If the input and output files are identical, how can the technique be considered a compression scheme?

Lossless JPEG really is lossless. However, as we just predicted, it fails to achieve compression ratios anywhere near those attainable by traditional JPEG, typically achieving ratios of 2:1 compared to traditional baseline JPEG ratios of as much as 40:1. JPEG-LS, sometimes called LOCO, achieves higher levels of compression than the original lossless JPEG, but it still falls short of traditional lossy JPEG.

JPEG, then, actually refers to a family of compression algorithms and does not specify a specific image format. In fact, when JPEG was originally being written, the JPEG committee was not allowed to define a specific file format because of disagreements within the international standards organizations. To get around this, a number of related standards, aimed at becoming workarounds, have been created. These include the *JPEG File Interchange Format* (JFIF), a relatively primitive and unintelligent format that simply serves as a pixel-transport scheme, and TIFF/JPEG, an extension of the original Aldus TIFF format. TIFF is a complex format that enables a user to record a great deal of information about an image. JFIF has become one of the preferred Internet image compression techniques and is commonly recorded as a JPEG file.

Apple's QuickTime software uses a JFIF-compatible data format that is encapsulated within the Macintosh-specific PICT format. Conversion between JFIF and PICT/JPEG is a simple process, and a number of Macintosh conversion applications are available.

The Graphics Interchange Format (GIF)   GIF defines a protocol designed to transmit raster-based graphic data that is independent of the hardware used to create or display it. GIF is defined in terms of blocks and subblocks that contain relevant information used to reproduce a graphic.

CompuServe released GIF as a free and open specification in 1987. It soon became a global standard and also played an important role within the Internet community as users started to share graphics files. It was well supported by CompuServe's Information Service, but did not require a subscription to CompuServe. The format was relatively simple and well documented.

Like most graphic management tools, GIF compresses images to reduce the file size, using a technique called *Lempel-Ziv-Welch* (LZW) compression. Unisys holds a patent on the LZW procedure, which soon became a popular technique for data compression. GIF is not the only tool that relies on LZW; TIFF, described in the following section, also includes LZW compression among its favored compression techniques.

JPEG versus GIF JPEG can typically achieve compression ratios of 10:1 to 20:1 without perceptible loss, 30:1 to 50:1 compression is possible with small to moderate visible artifacts, and 100:1 compression can be achieved, although quality suffers dramatically.

By comparison, a GIF image loses most of the color information in the process of reducing the 24-bit image to the 256-color palette, providing a 3:1 compression ratio. GIF's LZW compression scheme doesn't work well on photographs, yielding maximum compression levels of 5:1 and sometimes far less.

Because the human eye is more sensitive to luminance variations than it is to variations in color, JPEG compresses color data more than it compresses brightness data. Generally speaking, a grayscale image that is JPEG-encoded only achieves a 10 to 25 percent reduction of a full-color JPEG file of similar quality. The uncompressed grayscale image comprises 8 bits per pixel, or roughly a third of the color data. As a result, the compression ratio is much lower.

Tag Image File Format (TIFF) TIFF is a tag-based image format designed to promote the interoperability of digital images. The format came into being in 1986 when Aldus Corporation, working with leading scanner vendors, created a standard file format for images to be used in desktop publishing applications. The first version of the specification was published in July 1986; the most current version of the specification Version 6.0, was completed in September 1995 and is available on Adobe's web site at http://partners.adobe.com/asn/developer/PDFS/TN/TIFF6.pdf.

The format that defines a file specifies both the structure of the file and its content. TIFF content consists of a series of definitions of individual fields. The structure, on the other hand, describes how to actually find the fields. These pointers are called *tags*.

TIFF provides a general-purpose format that is compatible with a variety of scanners and image-processing applications. It is device independent and is acceptable to most operating systems, including Windows, Macintosh, and UNIX. The standard has been integrated into most scanner manufacturers' software and desktop publishing applications.

Adobe continues to enhance TIFF within publishing applications and maintains backward compatibility whenever possible.

Compressing Moving Images

Growth in videoconferencing, on-demand training, and gaming is fueling the growth in digital video technology, but the problems mentioned previously still loom large. Recent advances have had an impact; for example, storage and transport limitations can often be overcome with compression.

The most widely used compression standard for video is MPEG, which was created by the Moving Pictures Expert Group, the joint ISO/IEC/ITU-T organization that oversees standards development for video. MPEG is relatively straightforward. There are three types of MPEG frames created during the compression sequence. They are *intra* (I) frames, *predicted* (P) frames, and *bidirectional* (B) frames. An I-frame is nothing more than a frame that is coded as a still image and used as a reference frame. P-frames, on the other hand, are predicted from the most recently reconstructed I- or P-frame. B-frames are predicted from the closest two I- or P-frames, one from the past and one in the future.

Imagine the following scenario. You are converting a piece of video that you shot at the beach to MPEG. The scene, shown in Figure 1-78, lasts six seconds and is nothing more than footage of the fishing boat moving slowly in front of the camera, which is locked down on a tripod. Remember that video captures a series of still frames, one every one-thirtieth of a second (30 frames per second). What MPEG does is an analysis of the video based on the reference I-frames, P-frames, and B-frames. From the image shown in the illustration, it should be clear that very little changes from one frame to another in one-thirtieth of a second. The boat may move slightly (but very slightly), and the foam that the propeller is churning up will change. Other than that, very little in the scene changes. Without going into too much technical detail, what MPEG does is reuse those elements of the I-frame that don't change—or that change infrequently—so that it does not have to recreate them, thus reducing overall compression time. In our fishing boat scene, it should be

Figure 1-78
This image, taken from an MPEG stream, has few elements that change appreciably from frame to frame.

fairly obvious that the background certainly won't change much (unless a bird flies into it), the immediate foreground won't change, and the color and shape of the boat are constant. This is a fairly predictable scene. As a result, the number of I-frames that will be interspersed among the P-frames and B-frames is relatively small. If the scene were different—a fishing boat being tossed about on rough seas, for example—then the number of minimally uncompressed reference frames would be greater because of the constantly changing point of reference. MPEG looks backward to establish patterns of behavior of the past frames, looks at the reference frame, and finally predicts what future frames will probably look like based on past history. Ultimately, the sequence of frames is as follows: IBBPBBPBBPBBIBBPBBPB . . .

There are 12 frames from I to I, and the ratio of P-frames to B-frames is based on experience. Sequentially, the frames will look like the sequence shown in Figure 1-79.

The MPEG Standards There are several different versions of MPEG compression. They are described in the following sections.

I   B   B   P   B   B   P   B   B   P   B   B   I

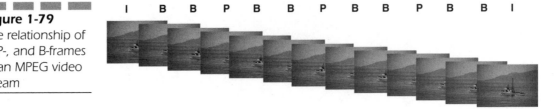

Figure 1-79

The relationship of I-, P-, and B-frames in an MPEG video stream

MPEG-1   MPEG-1 was created to solve the transmission and storage challenges associated with relatively low-bandwidth situations, such as PC-to-CD-ROM or low-bit-rate data circuits. MPEG-2, on the other hand, is designed to address much more sophisticated transmission schemes in the 6 to 40 Mbps range. This positions it to handle such applications as broadcast television and HDTV, as well as VBR video delivered over packet, Ethernet, and Token Ring networks.

MPEG-2   The original MPEG-1 standard was optimized for CD-ROM applications at 1.5 Mbps. Video, one of the applications that users wanted to place on CD, was a strictly noninterlaced format (progressive scan). Unfortunately, today's television pattern is interlaced. This created a problem; luckily MPEG-2 handles the conversion well.

MPEG-2 initially gained the attention of both the telephone and the cable television industries as the near-term facilitator of VOD services. Today, it is widely used in DVD production as well as in cable television, DBS, and HDTV because it has the capability to achieve high compression levels without the loss of quality. MPEG-2 has also found a home in corporate videos and training products because of its capability to achieve CD-compatible compression levels.

MPEG-3   MPEG-3, also known as MP3, has become both famous and infamous of late and has made the terms Napster and Metallica household names. MPEG layer 3 is a type of audio codec that delivers significant compression, as much as 12:1 of an audio signal with very little degradation of sound quality. Higher compression levels can be achieved, but sound quality suffers.

The standard bit rates used in MP3 CODECs are 128 and 112 Kbps. One advantage of MP3 is that a sound file can be broken into pieces and each piece remains independently playable. The feature that makes this possible means that MP3 files can be streamed across the net in real time, although network latency can limit the quality of the received signal.

The only real disadvantage of MP3 compression is that significant processor power is required to encode and play files. Dedicated MP3 players have recently emerged on the market and are proving to be quite popular. I have an MP3 player with 6GB of storage; I have encoded and loaded more than 170 CDs at CD quality into the thing and still have over a gigabyte of space left.

MPEG-4   MPEG-4 is a specially designed standard for lower-bit-rate transmissions such as dial-up video. MPEG-4 is the result of an international effort that involves hundreds of researchers from all over the world. Its development was finalized in October 1998 and became an International Standard in the first months of 1999.

MPEG-4 builds on the proven success of three fields: digital television, interactive graphics applications, and interactive multimedia such as the distribution of HTML-encoded web pages.

MPEG-7   MPEG-7, officially known as the *Multimedia Content Description Interface*, is an evolving standard used to describe multimedia content data that allows some degree of interpretation of the information's actual meaning, which can be accessed by a computer. MPEG-7 does not target any particular application; instead, it attempts to standardize the interpretation of image elements so that it can support a broad range of applications.

Consider the following scenario. A video editor, attempting to assemble a final cut of a video, finds that she needs a specific piece of B-roll (fill-in footage) to serve as a segue and to cover a voice-over during one particular scene in the video. She knows that the footage she needs exists in the company's film vault; she just can't remember what project it is filed under. Using MPEG-7, she requests the following: "I need a seven-second clip that has the following characteristics:

- A bright blue sky, midday
- A white beach in foreground, light surf
- A red and green beach umbrella in the lower-left corner of shot

- A seagull entering from upper left and exits to upper right
- A ship on horizon moving slowly from right to left"

With MPEG-7, the clip can be found using these descriptors.

The elements that MPEG-7 attempts to standardize support a broad range of applications such as digital libraries, the selection of broadcast media, postproduction and multimedia editing, and home entertainment devices. MPEG-7 will also add a new level of search capability to the Web, making it possible to search for specific multimedia content as easily as text can be searched for today. This capability will be particularly valuable to managers and users of large content archives and multimedia catalogs that enable people to select content for purchase. The same information used for successful content retrieval will also be used to select and filter content for highly targeted advertising applications.

Any domain that relies on the use of multimedia content will benefit from the deployment of MPEG-7. Consider the following list of examples, culled from the MPEG-7 standard located at http://mpeg.telecomitalialab.com/standards/mpeg-7/mpeg-7.htm#a_Toc533998968:

- Architecture, real estate, and interior design (for example, searching for ideas)
- Broadcast media selection (for example, radio channel and TV channel)
- Cultural services (for example, history museums, art galleries, and so on)
- Digital libraries (for example, image catalog, musical dictionary, biomedical-imaging catalogs, and film, video, and radio archives)
- E-commerce (for example, personalized advertising, online catalogs, and directories of e-shops)
- Education (for example, repositories of multimedia courses and multimedia searches for support material)
- Home entertainment (for example, systems for the management of personal multimedia collections, including the manipulation of content, such as home video editing, searching a game, and karaoke)
- Investigation services (for example, human characteristics for recognition and forensics)

- Journalism (for example, searching speeches of a certain politician using his or her name, voice, or face)

- Multimedia directory services (for example, yellow pages, tourist information, and geographical information systems)

- Multimedia editing (for example, personalized electronic news service and media authoring)

- Remote sensing (for example, cartography, ecology, and natural resources management)

- Shopping (for example, searching for clothes that you like)

- Social (for example, dating services)

- Surveillance (for example, traffic control, surface transportation, and nondestructive testing in hostile environments)

The standard also lists the following examples of how the capabilities of MPEG-7 might be used:

- A musician plays a few notes on a keyboard and retrieves a list of musical pieces similar to the required tune or images that match the notes in a certain way, as described by the artist.

- An artist draws a few lines on a screen and finds a set of images containing similar graphics, logos, and ideograms.

- A clothing designer selects objects, including color patches or textures and retrieves examples among which you select the interesting objects to compose your design.

- Using an excerpt of Pavarotti's voice, an opera fan retrieves a list of his records and video clips in which he sings a particular piece, as well as photographs of Pavarotti.

This is a remarkable (and remarkably complex) standard that holds enormous promise.

Other techniques exist, but they are largely proprietary. These include PLV, Indeo, RTV, Compact Video, AVC, and a few others.

QoS Issues

QoS is a matter of serious concern for both network providers and users because more than ever it defines the relationship that must exist between the two. QoS is important at all levels of the network hierarchy, from access through the transport network end to end. In this next section, we discuss the protocols and technologies that govern the delivery of QoS in modern networks, with an emphasis on video issues.

H.323 H.323 started as H.320 in 1996, an ITU-T standard for the transmission of multimedia content over ISDN. Its original goal was to connect LAN-based multimedia systems such as videoconferencing units to network-based multimedia systems. It defined a network architecture that included gatekeepers, which performed zone management and address conversion; endpoints, which were terminals and gateway devices; and multimedia control units, which served as bridges between multimedia types.

H.323 has been rolled out in four phases. Phase one defined a three-stage call setup process: a precall step, which performed user registration, connection admission, and the exchange of status messages required for call setup; the actual call setup process, which used messages similar to ISDN's Q.931; and, finally, a capability exchange stage, which established a logical communications channel between the communicating devices and identified conference management details.

Phase two allowed for the use of the RTP over ATM, which eliminated the added redundancy of IP and also provided for privacy, authentication, and greatly demanded telephony features such as call transfer and call forwarding. RTP has an added advantage. When errors result in packet loss, RTP does not request resends of those packets, thus providing for real-time processing of application-related content. No delays result from errors.

Phase three added the ability to transmit real-time fax after establishing a voice connection, and phase four, released in May 1999, added call connection over UDP, which significantly reduced call setup time, interzone communications, call hold, call park, call pickup, and call and message waiting features. This last phase bridged the considerable gap between IP voice and IP telephony.

One concern with H.323 is interoperability. There are currently three versions available, and because of a lack of deployment coordination, all three have a significant installed base that naturally leads to a certain amount of fragmentation. Consequently, interoperability has become a serious concern. One solution that has been offered is the iNOW! profile. Developed initially by Lucent, Vocal-Tec, and ITXC, iNOW! was created to define the options that manufacturers can use to ensure interoperability for multimedia voice and fax. INOW! recently merged with the *International Multimedia Teleconferencing Consortium* (IMTC) whose stated goals are to sponsor and conduct interoperability tests between suppliers of conferencing products and services, provide a forum for technical exchanges between IMTC members that will lead them to make additional submissions to the standards bodies to support interoperability and the usability of multimedia teleconferencing products and services, and, finally, to educate the business and consumer communities on the benefits of the underlying technologies and applications. Under the auspices of IMTC, iNOW! has the following responsibilities:

- To help members build interoperable IP telephony products based on the current and future IMTC iNOW! interoperability profiles developed and approved by the Activity Group.

- To provide a forum for technical exchange and to help with the resolution of interoperability technical issues that affect interoperability between iNOW!-compliant products and other products.

- To define test strategies that member companies can use to test their products for interoperability.

- To support IMTC and ETSI TIPHON testing of cross-vendor product testing.

- To support the development of additional features and a broader scope for IMTC iNOW! profiles as the need arises.

Several Internet interoperability concerns are addressed by H.323. These include gateway-to-gateway interoperability, which ensures that telephony can be accomplished between different vendors' gateways; gatekeeper-to-gatekeeper interoperability, which does the same thing for different vendors' gatekeeper devices; and finally gateway-to-gatekeeper interoperability, which completes the interoperability picture.

H.261 In 1984, after perceiving the need for a standard that would define video services based on ISDN, the *International Telegraph and Telephone Consultative Committee* (CCITT) (now the ITU-TSS) Study Group XV formed a group on Coding for Visual Telephony with the objective of recommending a standard for coded video transmission at P × 384 Kbps. Later it became clear that a single standard could cover the entire ISDN channel capacity. After five years of effort, CCITT Recommendation H.261, "Video Codec for Audiovisual Services at P × 64 Kbps," was completed and approved in December of 1990. A modified version was adopted for North America.

The applications of H.261 are video telephony and videoconferencing. Therefore, the recommended encoding algorithm must operate in real time with minimum delay. When P = 1 or 2, desktop videoconferencing is appropriate because of limited bandwidth. More complex images can be accommodated with reasonable quality when additional channels are added.

H.261, although still very much in existence, is fading in the presence of H.323.

Session Initialization Protocol (SIP) Although H.323 has its share of supporters, it is under attack from the IETF's *Session Initialization Protocol* (SIP). SIP supporters claim that H.323 is far too complex and rigid to serve as a standard for basic telephony setup requirements, arguing that SIP, which is architecturally simpler and imminently extensible, is a better choice. In reality, H.323 is an umbrella standard that includes (among others) H.225 for call handling, H.245 for call control, G.711 and G.721 for CODEC definitions, and T.120 for data conferencing. Originally created as a technique for transporting multimedia traffic over a LAN, gatekeeper functions have been added that allow LAN traffic and LAN capacity to be monitored so that calls are established only if adequate capacity is available on the network. Later, the Gatekeeper Routed Model was added, which allowed the gatekeeper to play an active role in the actual call setup process. This meant that H.323 had migrated from being a purely peer-to-peer protocol to having a more traditional, hierarchical design.

The greatest advantage that H.323 offers is maturity. It has been available for some time now, and although it is robust and full featured, it was not originally designed to serve as a peer-to-peer proto-col. Its maturity, therefore, may not be enough to carry it. It currently lacks NNI and does not adequately support congestion control. This is not generally a problem for private networks, but it does become problematic for service providers who want to interconnect PSTNs and provide national service among a cluster of providers. As a result of this, many service providers have chosen to deploy SIP instead of H.323 in their national networks.

SIP is designed to establish peer-to-peer sessions between Inter-net routers. The protocol defines a variety of server types, including feature servers, registration servers, and redirect servers. SIP sup-ports fully distributed services that reside in the actual user devices, and because it is based on existing IETF protocols, it provides a seamless integration path for voice/data integration.

Similar to H.323, SIP does not yet offer NNI, but the IETF has cre-ated a working group designed to bring together the best features of SIP and MeGaCo (an important protocol that will be discussed in detail in the next section) to overcome this obstacle.

Ultimately, telecommunications, like any industry, revolves around profitability. Any protocol that allows new services to be deployed inexpensively and quickly immediately catches the eye of service providers. Like TCP/IP, SIP provides an open architecture that can be used by any vendor to develop products, thus ensuring multivendor interoperability. Because SIP has been adopted by pow-erhouses such as Lucent, Nortel, Cisco, Ericsson, and 3Com and is designed for use in large carrier networks with potentially millions of ports, its success is reasonably assured.

Originally, H.323 was to be the protocol of choice to make this pos-sible. Although H.323 is clearly a capable suite of protocols and is indeed quite good for VoIP services that derive from ISDN imple-mentations, it is still incomplete and is quite complex. As a result, it has been relegated to use as a video control protocol and for some gatekeeper-to-gatekeeper communications functions.

The intense interest in moving voice to an IP infrastructure is dri-ven by simple and understandable factors: cost of service and enhanced flexibility. However, in keeping with the "Jurassic Park Effect" (just because you *can* doesn't necessarily mean you *should*), it is critical to understand the differences that exist between simple voice and full-blown telephony with its many enhanced features. It is the feature set that gives voice its range of capabilities. A typical local switch such as Lucent Technologies' 5ESS offers more than 3,000 features, and more will certainly follow. Of course, the features and services are possible because of the protocols that have been developed to provide them across an IP infrastructure.

Media Gateway Control Protocol (MGCP) and Friends Many of the protocols that are guiding the successful development of VoIP efforts today stem from work performed early on by Level 3 and Telcordia, which together founded an organization called the International SoftSwitch Consortium. In 1998, Level 3 brought together a collection of vendors who collaboratively developed and released the *Internet Protocol Device Control* (IPDC). At the same time, Telcordia created and released the *Simple Gateway Control Protocol* (SGCP). The two were later merged to form the *Media Gateway Control Protocol* (MGCP), which is discussed in detail in RFC 2705.

MGCP enables a network device responsible for establishing calls to control the devices that actually perform IP voice streaming. It permits software call agents and media gateway controllers to control streaming media gateways at the edge of the network. These gateways can be cable modems, set-top boxes, PBXs, VTOA gateways, and VoIP gateways. Under this design, the gateways manage the circuit-switch-to-IP voice conversion, while the agents manage signaling and call processing.

MGCP makes the assumption that call control in the network is software based, which is resident in external intelligent devices that perform all call control functions. It also makes the assumption that these devices will communicate with one another in a primary-secondary arrangement, under which the call agents send instructions to the gateways for execution.

Meanwhile, Lucent created a new protocol called the *Media Device Control Protocol* (MDCP). The best features of the original three were combined to create a full-featured protocol called the MeGaCo, which was also defined as H.248. In March 1999, the IETF and ITU met collaboratively and created a formal technical agreement between the two organizations, which resulted in a single protocol with two names. The IETF calls it MeGaCo; the ITU calls it H.GCP.

MeGaCo/H.GCP operates under the assumption that network intelligence is housed in the central office and therefore replaces the gatekeeper concept proposed by H.323. By managing multiple gateways within a single IP-equipped central office, MeGaCo minimizes the complexity of the telephone network. In other words, a corporation might be connected to an IP-capable central office, but because of the IP-capable switches in the central office, which have the capability to convert between circuit- and packet-switched voice, full telephony features are possible. Thus, the next-generation switch converts between circuit and packet, while MeGaCo performs the signaling necessary to establish a call across an IP WAN. It effectively bridges the gap between legacy SS7 signaling and the new requirements of IP, and supports both connection-oriented and connectionless services.

As Table 1-4 shows, bandwidth requirements vary from application to application. Now that we have transported the video signal across the network, we must terminate it with a viewing device such as a television, PC, or videoconferencing unit. The section that follows discusses the issues of signal termination.

Destination Signal Issues

It's interesting to note that in the drive toward digitization the ultimate goal is to create a video storage and transport technology that

Table 1-4

Bandwidth required for typical applications

Application	Required Bandwidth	Sensitivity to Delay
Voice	Low	High
Video	High	Medium to high
Medical imaging	Medium to high	Low
Web surfing	Medium	Medium
LAN interconnection	Low to high	Low
E-mail	Low	Low

will yield as good of a representation of the original image as analog transmission does.

There are two primary governing organizations that dictate television standards. One is the National Television System Committee (NTSC) (sometimes said to stand for *Never Twice the Same Color* based on the sloppy color management that characterizes the standard); the other, used primarily in Europe, is the Phased Alternate Line (PAL) system. NTSC is built around a 525-lines-per-frame and 30-frames-per-second standard, whereas PAL uses 625-lines-per-frame and 25-frames-per-second. Although technically different, both address the same concerns and rely on the same characteristics to guarantee image quality. A third standard, *Sequential Couleur avec Mémoire* (SECAM), is used in a variety of locations around the world.

Video Quality Factors   Four factors influence the richness of the video signal. They are the frame rate, color resolution, image quality, and spatial resolution.

Frame Rate   Frame rate is a measure of the refresh rate of the actual image painted on the screen. The NTSC video standard is 30 frames per second, meaning that the image is updated 30 times every second. Each frame consists of odd and even fields. The odd field contains the odd-numbered screen lines, while the even field contains the even-numbered screen lines that make up the picture.

Television sets paint the screen by first painting the odd field, and then the even. They repeat this process at the rate of 60 fields—or 30 frames—per second. The number 60 is chosen to coincide with the frequency of electricity in the United States. By the same token, PAL relies on a scan rate that is very close to 50 Hz, the standard in Europe.

Video can be displayed using two distinct formats: *progressive scan* and *interlaced scan*. Progressive scan, most commonly seen in computer monitors and digital television displays, paints all of the horizontal lines of the TV image in a single pass, creating a single *frame*. On the other hand, interlaced scan, commonly used in standard television formats such as NTSC, PAL, and SECAM, displays half of the horizontal lines at a time. It first paints the odd-numbered lines, known as a *field*, followed by the even-numbered lines to create a second field. The fields, when interlaced, create a single frame. This is illustrated in Figure 1-80.

An interlaced scan takes advantage of a phenomenon called *phosphor persistence*. The phosphorescent compound that is painted on the inside surface of the front of the TV tube and that is excited by the high-energy sweep beam that strikes it continues to glow faintly after it has been painted. As a result of the latency of the human eye, the two fields blend together to create what appears to be a single picture.

The advantage of interlaced video is that a high image refresh rate (50 Hz in Europe and 60 Hz in the United States) can be achieved with half the bandwidth. The disadvantage is that the horizontal resolution of the screen image is halved, which can result in a flickering screen and other visual artifacts.

Figure 1-80
Two fields make up a frame in interlaced video.

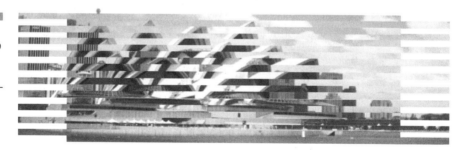

An Aside: DVD Because of its popularity, we should note that DVD is specifically designed to be displayed on interlaced-scan displays, which covers virtually 100 percent of all televisions in operation today. However, most DVD content comes from film, which is created using progressive scanning techniques. To make film content work in interlaced form, the video from each film frame is split into two video fields—240 lines in one field and 240 lines in the other—and encoded separately in an MPEG-2 stream. One challenge that must be dealt with is that film runs at 24 frames per second, while TV runs at 30 frames per second (60 fields) in NTSC systems, or 25 frames per second (50 fields) for PAL and SECAM systems. For PAL and SECAM, the easiest way around this apparent disparity is to show the film frames at 25 per second and accelerate the audio accordingly. For NTSC, the common solution is to match 24 film frames to 60 video fields by alternating the display of the first film frame for 2 video fields and the next film frame for 3 video fields, which is a technique known as a *2–3 pulldown*.

Because DVD often uses MPEG-2 compression, the repeated fields are not actually stored twice. Instead, a flag in the protocol is set that tells the image decoder to simply repeat the field.

Of course, some problems emerge with this technique. For example, some frames in a film are shown for a longer period of time than others, which causes as a jerky image during slow camera pans.

DVD can reproduce video and audio that are better than broadcast video and CD-quality audio. DVD video is usually encoded from the master tapes using MPEG-2 compression. As we discussed earlier, this process eliminates redundant information in the series of images because of the known limitations of the human eye. The resulting video, particularly when it is characterized by a lot of movement and action, may contain visual flaws, depending on the degree of compression used to create it. At typical data rates of 3.5 to 5 Mbps, compression artifacts may occasionally appear. Higher bandwidth can result in higher quality, and as MPEG compression technology improves, these artifacts will become a thing of the past.

There are two primary display formats for video today, both of which are commonly seen on DVDs. Video can be stored on a DVD in what is called *4:3 format*, which creates the standard shape of a typical TV screen, or *16:9 format*, which is commonly called *widescreen* or *letterbox* format. The width-to-height ratio of standard television displays is 4 to 3, meaning that it is 1.33 times wider than it is tall. The widescreen television sets that have only recently become available and that are targeted at the HDTV market have a ratio of 16 to 9. DVD is designed to support widescreen displays. (You know this; many DVDs ask what format you would like to use when you start a movie.)

Letterbox format implies that the video is displayed in the aspect ratio used in most theaters, a format that is wider than standard or widescreen television. Horizontal black bars, called *mattes*, cover the image gaps at the top and bottom of the screen.

DVDs compare quite favorably to CDs in terms of capacity and quality. A typical CD can store about 650MB, while a single-side, single-layer DVD holds nearly 5GB of data. For comparison purposes, approximately 2GB of disk space is required for one hour of video.

Table 1-5, courtesy of Jim Taylor's *DVD Demystified FAQ* located at http://www.dvddemystified.com/dvdfaq.html, shows the typical capacity of common DVD formats.

Computers often rely on *Variable Graphics Array* (VGA) monitors that are much sharper and clearer than television screens. This is due to the density of the phosphor dots on the inside of the screen face that yield color when struck by the deflected electron beams, as well as a number of other factors. The scan rate of VGA is much higher than that of traditional television and can therefore be non-interlaced to reduce screen flicker. Alternatively, if a VGA signal is to

Table 1-5		
	DVD-5 (12 cm, SS/SL)	4.37 GB of data, over 2 hours of video
DVD formats	DVD-9 (12 cm, SS/DL)	7.95 GB, about 4 hours
	DVD-10 (12 cm, DS/SL)	8.74 GB, about 4.5 hours
	DVD-14 (12 cm, DS/ML)	12.32 GB, about 6.5 hours
	DVD-18 (12 cm, DS/DL)	15.90 GB, over 8 hours
	DVD-R 1.0 (12 cm, SS/SL)	3.68 GB
	DVD-RW 2.0 (12 cm, SS/SL)	4.37 GB; 8.75 gig for rare DS discs

be displayed on a TV monitor, a scan converter, which was described earlier, can convert the signal as required for display.

Color Resolution

Another quality factor is color resolution. Most systems resolve color images using a technique called RGB—for the red, green, and blue primary colors. While video does rely on RGB, it also uses a variety of other resolution techniques.

Image Quality

Image quality plays a critical role in the final outcome, and the actual resolution varies by application. For example, the user of a slow-scan, desktop videoconferencing application might be perfectly happy with a half-screen, 15-frames-per-second, 8-bit image, whereas a physician using a medical application might require a full-frame, high-resolution image with 24-bit color for perfect accuracy. Both frame rate (frames per second) and color density (bits per pixel) play a key role.

Spatial Resolution

Finally, spatial resolution comes into the equation. Many PCs have displays that measure 640 $\times$ 480 pixels. This is considerably smaller than the NTSC standard of 768 $\times$ 484 or even the slightly different European PAL system. In modern systems, the user has great control over the resolution of the image because he or she can vary the number of pixels on the screen. The pixels on the screen are simply memory representations of the information displayed. By selecting more pixels, and therefore better resolution, the graininess of the screen image is reduced. Some VGA adapters, for example, have resolutions as dense as 1,024 $\times$ 768 pixels and higher.

The converse, of course, is also true. By selecting less pixels, and therefore increasing the graininess of the image, special effects can be created, such as pixelization or tiling.

One termination or delivery technology that we have not yet described is virtual reality. Once considered to be something of a vanity application, virtual reality has gained significant importance in a wide variety of industries.

Virtual Reality

Asking for a definition of *virtual reality* is reminiscent of the parable of the five blind men and the elephant in which a group of five blind men is asked to approach and examine an elephant and then return to describe the creature. One man approached and felt the elephant's leg, another felt its trunk, and the others felt its ear, tail, and broad side. Their descriptions of the elephant were all quite different.

Today, virtual reality has the interest of a fair number of companies and industries. Over the last few years, video has brought significant change to the business and academic community; today, virtual reality provides a new focus.

Virtual reality is different things to different people. To some, virtual reality is a computer application that provides a doorway into imaginary worlds. To others, it is the remarkable collection of hardware and software that provides the portal mentioned previously. Others escape reality through books, movies, and fantasy games such as Dungeons and Dragons—to them, that's another virtual reality. Whatever the case, virtual reality is a marriage of technology and human imagination that is insinuating itself into some remarkable—and perhaps surprising—applications.

So, what is virtual reality? One text on the subject defines it as *a way for humans to visualize, manipulate, and interact with computers and extremely complex data.* This definition actually describes virtual reality fairly well. Virtual reality relies on computer-generated visual and auditory sensory information about a "world" that only exists within the computer. This world could be a CAD drawing of a cathedral, a model showing chemicals interacting, or air flowing across the wings of an aircraft. Obviously, the data required to drive the virtual-reality engine is massively complex and requires sophisticated computing capability.

To some, virtual reality is something of an oxymoron (Mark Weiser of Xerox PARC, a pioneer in virtual reality, called it "real virtuality"). Indeed, early systems were not particularly convincing. The bandwidth and computer power required for the computer software to keep up with the user's complex movements were not up to the task, and there was typically a lot of latency.

Today, things are quite different. Not only are the graphics significantly cleaner, the computers and orders of magnitude faster, and the equipment sleek and comfortable, but some systems have now added a tactile element. Special gloves and suits are available that actually give the user tactile feedback in addition to sound and vision. Virtual surgery applications, for example, provide sensation to the surgeon, giving the actual feel of the real procedure. Virtual-reality body suits, using servos and microbladders that rapidly fill and deflate with air, can be tied into applications to yield a wide variety of (interesting) tactile sensations ranging from the punch of a boxer to—well, use your own imagination.

Virtual-Reality Techniques

Today, there are six distinct "flavors" of virtual reality. The first flavor, often called *Window on the World* (WoW), was one of the earliest developments and is certainly the least complex. In WoW, a standard computer screen displays two-dimensional application graphics and serves as a window into the virtual environment that the application creates. Adjacent speakers provide sound. WoW first emerged in early 1965.

The second technique is called *Video Mapping* and is a variation of WoW. In a Video Mapping system, the application overlays a video image of the user's silhouette on the graphics, and the user watches his or her own image as it wends its way through the virtual world. As with WoW systems, it is still a two-dimensional application, but is improved by the addition of the user.

Video Mapping emerged in the late 1960s. Several early commercial virtual-reality applications used Video Mapping, including Mandala, which relied on a modified Commodore Amiga, and the TV station Nickelodeon, which used a Video Mapping application on its *Nick Arcade* to overlay competitors on a giant video arcade game.

A third technique, and perhaps the most widely used, is called *immersive virtual reality*. Immersive systems typically use some kind of *head-mounted display* (HMD) to provide 3-D video and stereophonic sound, thus immersing the user in the world created by the application. The helmet also houses motion-detection devices, which tell the application where the user is, what direction he or she is looking, and what he or she should be seeing and hearing. The helmet is usually tethered by a cable back to the computer, although some new systems rely on wireless connectivity.

A variation of HMD-reliant systems is the so-called Cave, like the one conceived and created at the University of Illinois at Chicago's Electronic Visualization Laboratory. "Virtual Reality is the descendent of a long line of computer graphics research that tries to incorporate more modalities of perception and interaction into our interface with computers," says Dan Sandin, codirector of the lab. "Many people believe that VR systems have to use helmets and data gloves, which are actually somewhat restrictive. We created the Cave as a solution to the problem of reduced fields of vision that plague many of the HMD-type VR systems.

"The Cave is a projection-based virtual reality system, in which three-dimensional images are projected on all six sides (four walls, the ceiling, and the floor). The user wears a pair of lightweight polarized glasses that also house a very small, wireless motion and location sensor (they look like the nerdy black hornrims of the '60s). When the Cave is activated, the user suddenly finds him- or herself immersed in a seemingly endless school of sharks, or floating in space outside the Shuttle, or dancing with an indescribable, three-dimensional polygon. It's quite a remarkable thing."

Perhaps the best-known "virtual" virtual-reality application is the Holodeck used on *Star Trek: The Next Generation*. Sandin smiles at the comparison to the Cave. "The thing that differentiates the Cave from the Holodeck is that you can't sit in the chairs in the cave. They look pretty good, and they're there, but don't try to sit in them."

The fourth major application of virtual reality is called *virtual telepresence*. In these systems, the user's senses are electronically linked to remote sensors, giving the user the sensation of actually being—and feeling—where the remote sensors are. In remote surgery applications (mentioned previously), systems are designed to give physicians tactile feedback as they operate. Firefighters and bomb squad personnel use remote telepresence applications in very delicate—and potentially dangerous—procedures. NASA and the Woods Hole Oceanographic Institute (Outer Space and Inner Space) use telepresence applications for the robotic exploration of deep space and the deep ocean. In fact, there is already a joint U.S./Russian robotic space rover program underway.

The fifth flavor of virtual reality is called *mixed reality*. In mixed reality applications, real-world inputs are combined with computer-generated images to facilitate whatever task is being attempted. In medicine, for example, a surgeon's video image of the actual organ might be overlaid with a CAT scan image to help best target the procedure. An application mentioned earlier combines virtual-reality imagery with CAT scan images to create a 3-D view of the patient's internal organs.

In combat aviation, images of the ground are combined with terrain and elevation information, and then displayed on the pilot's heads-up display or the inside of the helmet visor.

The last form of virtual reality is called *fishtank virtual reality*. It is the newest technique and is perhaps the most promising. In fishtank systems, the user wears a pair of lightweight glasses, similar to the Polaroid glasses worn in the Cave. These, however, have color LCD lenses that provide wide-field stereoscopic vision and include a mechanical head tracker unit.

Virtual-Reality Equipment

There are six principle hardware components in a virtual reality system: the computer that generates the images, manipulation devices, position trackers, vision systems, sound drivers, and the HMD.

Computer complexity ranges from inexpensive PC systems to high-end servers costing well over $100,000. Obviously, the more complex the graphics being generated, the more capable and powerful the image generator needs to be.

Manipulation and control devices provide the means of tracking the position of a real object within a virtual world. Because the real world is 3-D, tracking devices must be able to respond to roll, pitch, and yaw movements.

There are many different versions of these devices. The simplest are mice, trackballs, and joysticks. They provide the least control over virtual motion. A step above them is the instrumented glove, in which the fingers are equipped with sensors that notify the computer of movement and hand action. These range in complexity from gloves that use fiber-optic sensors for finger motion and magnetic trackers that detect general position to less complex devices that provide limited sensing with strain gauges on the fingers and ultrasonic overall position detectors.

Today, full-body suits and electronic exoskeletons are used to capture full character motion for animation applications, control of musical synthesizers, and some advanced games. One airline in-flight catalog sells a $10,000 body suit that can be connected to existing video games, such as Nintendo, Sony, and Sega—for the person that has *virtually* everything.

Position trackers can be as simple as a mechanical arm that follows the user around the virtual environment or as complex as full-body exoskeletons like those mentioned previously. Some companies offer a body suit that not only provides position information to the application, but delivers force feedback to the wearer of the suit to simulate real interaction with objects in the virtual world (an area of research known as *haptics*). Logitech is a well-known manufacturer of these devices.

The actual tracking units rely on a variety of technologies. *Ultrasonic sensors* pulse at a known repetitive frequency, and the latency between pulse and pickup is used to track location—like sonar. Magnetic sensors use coils that produce magnetic fields that can be tracked by the pickup unit. The best known manufacturer of magnetic units is Polhemus in Colchester, Vermont.

A third tracking system relies on *optical pickups* to determine location. Using a grid of ceiling-mounted LEDs that flash in a known pattern, helmet-mounted video cameras can track the wearer's location under the grid. Origin Instruments is the primary manufacturer of these systems.

Finally, there are a limited number of *inertial trackers* that can be used in virtual-reality systems. They typically only detect rotational movement, but in some systems, this is adequate.

We've already discussed many of the *vision systems* used in virtual reality. These include rear-projection systems, such as those in the UIC Cave, dual screens placed before the eyes to yield stereoscopic vision, and a split screen technique, in which the user wears a hood and images are painted on the screen in two parts, thus simulating stereoscopic vision.

True immersive systems require some sort of HMD. These can be as complex as a helmet containing stereophonic sound speakers, stereoscopic vision screens, and complex motion detection devices, to the pair of nerdy black polarized glasses (mentioned earlier) with a small motion sensor connected via a fine optical cable. These devices are currently expensive ($3,000 to $10,000), but are expected to plummet in the next few years as entertainment companies announce virtual-reality-based video games. Both Nintendo and Sega have plans underway to create and release an entire line of virtual-reality-dependent games within the next year.

An Image of the Future: Back to *Real* Reality

So, where is all this image-based technology going? In *Snow Crash,* author Neal Stephenson describes what is perhaps the ultimate marriage of virtual reality and the Information Superhighway. In his world, users log onto the network through a low-power wireless terminal that sports a lightweight HMD about the size of a pair of glasses. Instead of using a VDT or LCD displays, the HMD uses low-power, multicolored lasers to paint 3-D images on the retinas of both eyes. The lasers have tracking circuitry that enables them to follow the minute movements of the user's eyes, thus guaranteeing a perfect set of images from any position.

Once logged on (or "jacked in," to use the appropriate terminology), the user finds him- or herself in a virtual world. The world—known as a *metaverse*—consists of a main street with shops, bars, and businesses (the databases of the more routine Web) that the user can visit. Furthermore, rather than talking to other users via the crude medium of typed text, users actually "see" other users in the virtual world. These manifestations of the real users are called *avatars*, a word that means "the physical manifestation of a god."

Obviously, we're quite a ways from this level of capability, but not all *that* far. Still-image applications are firmly entrenched and offer a broad variety of capabilities that have in many ways revolutionized banking, medicine, insurance, and architecture, to name a few. Video, too, has revolutionized many fields, but in certain tantalizing ways, technology has revolutionized video. Modern nonlinear systems digitally encode the video images captured on tape, store them on disk, and allow editors to manipulate the footage very easily. Home-based systems are now freely available and are well within the price range of the average hobbyist.

The exciting part of all this is the ultimate combination of virtual reality and digital video, high-bandwidth transport technologies, and emerging applications. Dan Sandin claims, "Not only will VR provide entertainment value, it will also become firmly entrenched in business. As businesses continue to realize how expensive and time consuming it is to move people around the country for meetings, and as the VR technology gets better and better, eventually there will be this mass realization that virtual presence—called telepresence in the jargon—is good enough, and perhaps better, if you factor in travel costs and time away from work." Add the further enhancement of desktop video and the future is here.

The delivery mechanism for all these visually-oriented systems is equally complex and exciting. One area that is as yet undefined is the whole question of who the network players will be. There is a great jockeying for position going on between telephone companies, cable providers, and application designers. Already, alliances are emerging that will help determine the shape of the interactive, entertainment-oriented future.

And what about the nonentertainment market? There are indeed applications out there, built around image, video, and virtual reality. Medicine and aviation, for example, enhance safety through the use of these technologies, while chemistry and architecture use them to enhance the understanding of highly complex structures and interactions. In education, the virtual classroom delivers teachers, schoolrooms, and simulations to far-flung locations and audiences.

Today, the cost of these technologies is somewhat prohibitive because they are new and provided by a relatively small number of pioneer manufacturers. As applications and the market drive them toward commodity, the cost will come down and digital video and virtual reality will become mainstream technologies.

In this rather long section of the book, we have examined conferencing applications with an emphasis on video. We then examined signal origination, access, transport, and termination technologies. In Part Two, we turn our attention to the videoconferencing industry and the companies that comprise it.

The Videoconferencing Industry

In this section we dissect the videoconferencing industry, which comprises four major segments as shown in Figure 2-1: the system manufacturers, the bridge providers, the backbone providers, and the videoconferencing service providers.

This industry as a whole is changing as it comes to grips with its newfound importance. There was a time, for example, when audio conferencing and videoconferencing were separate and unrelated, both technologically and in terms of the applications to which they were applied. That is no longer the case. Today, combined with webcasting, web conferencing, data conferencing, and media streaming, a new breed of collaborative interaction is coming into existence. There is finally enough bandwidth in the access networks to make these multimedia applications viable. Some of them are fully interactive, some one-way only, and some hybrids of multiple technologies. Additionally, they are not all designed to be interactive in real-time. There is a growing demand for streamed content from an archival server, which viewers can download on demand. The conferencing marketplace is now comprised of a variety of innovative applications that are to a very large extent technology agnostic. Some connect using ISDN, others via the PC, and some will connect using nothing more complex than a telephone. The key is that these

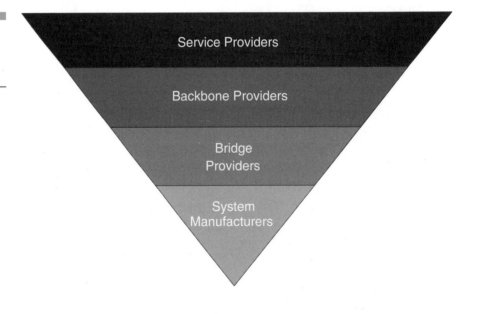

Figure 2-1
The video-
conferencing
industry

Service Providers

Backbone Providers

Bridge
Providers

System
Manufacturers

technologies now adequately satisfy a broad array of customer requirements, and that, more than anything else, is driving the popularity of conferencing solutions.

One new arrival on the videoconferencing scene is the so-called *Content Service Provider* (CSP). These companies offer the entire range of delivery modes including web-based service, audio conferencing, streaming media, and videoconferencing. Some of them are trying to move into IP-based services, while others remain with the time-tested capabilities of traditional circuit-switched networks such as the PSTN, frame relay, and ATM.

If nothing else, CSPs are performing one service for the industry as a whole that is invaluable: they are driving home the message that service counts more than anything else. Customers want a single provider that they can call for all of their conferencing requirements. They want to be able to call a single person who will help them schedule conferencing resources, reserve a public room if necessary, train their people, help them select equipment, perform troubleshooting, hold their hand, record the conference if requested, and be there when the conference takes place to ensure that everything comes off without a hitch. The bandwidth providers, bridge providers, and hardware manufacturers are rapidly approaching commodity status in this business, and the smart ones see it and are taking steps to forestall the impact of the inevitable. For example, they are investing in value-added services as a way to bolster revenues and provide differentiation. They are partnering with content providers and full-service videoconferencing providers to create the all-inclusive package that customers are looking for.

System Manufacturers

This section provides an overview of the manufacturers who build the videoconferencing units, bridges, and other devices that enable the videoconferencing marketplace. This is not a complete list; it includes descriptions and web addresses of the major players. Please visit their web sites for information about specific products. Thanks to all who provided content for this section of the book.

Accord Networks

Accord Networks produces next-generation network solutions that address a variety of network issues, including variable bandwidth and most existing and emerging standards. Accord manufactures, markets, and supports the MGC-100, a *Multipoint Control Unit* (MCU), and Gateway that enables live, interactive audio, video, and data communication between two or more conferencing systems (www.accordnetworks.com).

Aethra (Italy)

Aethra manufactures a broad array of videoconferencing products. The company's service center provides complete support for all products, including 24-hour technical support via Internet, fax, and telephone. The company's products allow a customer to conduct interactive multimedia sessions with multiple sites equipped with an ITU standard-compliant videoconference system over ISDN. Each site can view the other participants and work with them in real time (www.aethra.com).

Compunetix, Inc.

Compunetix provides multipoint multimedia telecommunications systems for both commercial and government markets. To better support its global customers, Compunetix has offices in Australia, the United Kingdom, and Hong Kong.

The *Video Systems Division* (VSD) of Compunetix is dedicated to the design, sales, and support of advanced multimedia, multipoint teleconferencing systems that expand the boundaries of modern business communications (www.compunetix.com).

First Virtual Communications

First Virtual Communications provides easy-to-use, integrated communications solutions to enterprises, service providers, and portals.

By enabling interactive voice, video, and data applications over IP, First Virtual provides cost-effective, integrated solutions on an end-to-end basis for everything from large-scale enterprise deployment to single desktops. It also supports videoconferencing solutions over ISDN and ATM networks (www.cuseeme.com).

The company's best-known product, Click to Meet™, offers a framework for delivering video-enabled web collaboration applications that address the communications requirements of client companies. Click to Meet can be integrated seamlessly into enterprise messaging and collaboration environments such as Microsoft Exchange/Outlook and instant messaging. The company serves its global enterprise, service provider, and portal customers through a worldwide network of resellers and partners.

First Virtual's Conference Server creates a "virtual conference room" on existing IP networks using H.320, H.323, SIP, and T.120 conferencing standards.

The Conference Server features

- A wide range of endpoints, supported at varying bandwidth rates

- Multiple server configurations, including Windows NT, Windows 2000, Sun Solaris, and Linux operating systems

- Diverse applications

- Flexible topology customization

- Easy scalability

- No limit on the number of linked or cascaded conferences for seamless scalability

- Transparent access as conferences span multiple servers

- No limits on the number of continuous conferences enabled

- A user-friendly interface that lets participants set up conferences at any time, from anywhere, via their browser

The company's bandwidth management and security control features include

- A built-in gatekeeper that regulates bandwidth used by H.323 clients

- Multicast capabilities
- Graphical server topology for large-scale deployment configuration
- Simultaneous seamless access by low and high bandwidth users without compromising any user's conference experience

Gentner

Gentner's expertise in audio conferencing technology is well known. The majority of their installations, however, are videoconferencing applications. The company's increased focus on videoconferencing products has yielded a high-quality line of products (www.gentner.com).

PictureTel

Acquired by PolyCom in 1991, PictureTel's products are designed to meet a wide range of customer needs from "head-and-shoulders" videoconferencing to advanced multisite collaboration based on extensive information sharing. The company offers products designed for both large and small enterprise groups, as well as products for individuals at their desks or at home. PictureTel's multipoint network systems architecture is designed to support users connected by multiple networks.

Larger group systems handle the most demanding applications in a conference room or auditorium situation. Other group systems are designed for situations that require high quality, power, and portability. Personal videoconferencing systems bring videoconferencing capability directly to an individual user's PC (www.picturetel.com).

Polycom

Polycom Inc., known worldwide for high-quality, easy-to-use communications equipment, enables businesses to access broadband network services and take advantage of high bandwidth to conduct video, voice, and data communications. The company's products include group videoconferencing systems, interactive video communications appliances for high-quality video calls from a desktop or laptop computer, full-duplex audio conferencing phones that minimize echo, clipping, and distortion, and DSL-based Integrated Access Devices (www.polycom.com).

RADVision

RADVision specializes in real-time IP communications for next generation networks. Their products, used by hundreds of companies, enable the Internet infrastructure required to support real-time voice and video transport.

RADVision's line of V^2oIP products offers a complete suite of IP-based devices (www.radvision.com).

Sony

Sony manufactures both videoconferencing units and a variety of interactive multimedia devices that complement their videoconference product line, including full compatibility with existing Sony products (www.sony.com).

Spectel

Spectel manufactures multimedia conferencing solutions for service bureaus, next generation carriers, *Application Service Providers* (ASPs) and corporations worldwide (www.spectel.com). Spectel's solutions include

- Integrated voice and data collaboration
- Integrated audio and video streaming
- Single platform reserved and reservationless conferencing
- Single platform attended and unattended conferencing

Tandberg

Tandberg is a global leader in videoconferencing. The company designs, develops, and manufactures videoconferencing systems and offers both sales support and value-added services throughout the world. Headquartered in Norway, Tandberg has offices in the United States, the United Kingdom, Canada, China, and Japan (www. tandbergusa.com).

VBrick Systems

Incorporated in early 1998, VBrick Systems, Inc. develops, designs, manufactures, and supports hardware and software products that deliver both real-time and on-demand DVD-quality video and CD-quality audio.

VBrick is the market leader for simple, low-cost video products that deliver one-way and two-way television over broadband networks. The company sells video solutions through a global network of partners in vertical markets that include education, security and monitoring, broadcasting, and corporate communications (www.vbrick.com).

VCON

VCON develops, manufactures, and markets multifunction personal and group videoconferencing systems for a variety of communication networks. The company's systems maximize performance over IP networks and support two-way videoconferencing and streaming of data and collaborative sharing of computer files. VCON's systems are primarily targeted at business, distance learning, government, and telemedicine applications. Their video-over-IP (VoIP) technology enables transmission and reception of video, voice, and data streams through existing networking infrastructures. Most systems provide a dual-mode capability, which permits them to be used in both IP and ISDN networks (www.vcon.com).

VTEL

VTEL Corporation is the world's largest developer and manufacturer of visual communication technology. Their products cover every environment, from the desktop to the boardroom, and incorporate both traditional data networks and the Internet. VTEL's solutions are easy to use, rely on mainstream architecture, are fully scaleable, provide for the management of large networks, and are customer focused. The company's solutions are interactive, pragmatic, and based exclusively on digital technology (www.vtel.com).

Zydacron

Zydacron builds "Intelligent Meeting Solutions" to help reduce the cost of meetings by increasing the effectiveness of meetings. The company's "Intelligent Meeting Solutions" integrate meeting and presentation tools to create a PC-based conferencing system that permits data sharing and site-to-site collaboration. Zydacron offers desktop systems, small group/small room systems, enterprise systems, and system integrators, which include CODECs and other component devices (www.zydacron.com).

Bridge Providers

This section provides an overview of the bridge providers, the companies that provide the "gathering place" that facilitates the establishment of a conference across the network. As with the system manufacturers, this list is far from complete, but gives a sense of the market and the companies that compose it. Please refer to these companies' web sites for additional information about their services.

AT&T Teleconference Services

AT&T offers a variety of audio and video conferencing services, either directly or through relationships with subsidiaries.

- *Audio Conferencing Services*   Can be toll free or caller paid. The customer can arrange the meeting, or an AT&T Operator can arrange it. If desired, the conference can be recorded digitally so that listeners can hear it later in a variety of formats and distribution methods.

- *Reservationless Service*   A subscription service that enables a customer to establish a meeting on demand.

- *Executive Offer*   Enables a customer to host press conferences or annual meetings across long distances.

- *Executive ConferenceCasting*   Uses the Internet to distribute PowerPoint or other presentation graphics to the audience while a facilitator discusses them live. The presentation can be recorded for on-demand listening and viewing.

- *Web Meeting Service*   Combines a conference call with data sharing capability over the Web. If required, the group can edit files, create sketches, share desktop control, chat among each other, survey participants online, and share the results of the survey immediately.

- *Videoconferencing*   Offers a multipoint video bridging service that enables three or more sites to join the same video conference call (www.att.com).

ACT Teleconferencing

ACT's suite of applications for audio, video, and Internet conferencing provides easy-to-use communications tools (www.acttel.com). The company's services include

- *ReadyConnect*   Audio conference calls without advance reservations or operator assistance.

- *ActionCall*   Gives the customer the ability to conduct a high-level audio conference call with full operator support.

- *Passcode*   Provides a secured audio conference, based on secure delivery of a passcode.

- *ActionView*   The capability for a one-to-many conference via video monitor to participants in worldwide offices.

- *ClarionCall*   Delivers full duplex unattended conferencing over your VoIP network.

- *ActionCast* and *ActionCast Plus*   Enables a client to stream an audio message over the Internet for an effective conference presentation.

- *ReadyConnect Online*   Delivers an "online tour guide," providing the capability to show presentations, guide participants through a web tour, annotate slides in real time, and share applications during a collaborative meeting.

Global Crossing

Global Crossing's conferencing solutions enhance the effectiveness of business communications. Service is critical to Global Crossing: They actively monitor each conference, troubleshoot as required, and constantly test video and audio quality (www.globalcrossing.com).

InView

InView was founded in 1995 as a spin-off of InterCall with a charter to provide videoconference bridging services. The company came about because of a growing demand for video services in addition to audio conferencing-based collaboration.

Since that time, InView has expanded into ISDN-based videoconference services, including web-based visual and audio collaboration solutions.

InView relies on the newest products from Accord, RADVision, and Ridgeway in its operations center to provide interoperable H.320/H.323 collaborative solutions and seamless communications between IP and ISDN networks (www.inview.com).

WorldCom

WorldCom Conferencing offers advanced conferencing services, including audio conferencing, with tape playback and transcription, and videoconferencing (www.worldcom.com).

- *Net Conferencing* Enables participants to view materials over the Internet via their PCs while they listen to the meeting on a conference call.
- *Conference Webcast* Integrates audio, video, and data communications.
- *Conferencing Event Services* Enables a customer to enhance live events with webcasts and conduct events entirely online.

V-Span

V-SPAN services include

- Dial-out/dial-in
- Multiple bandwidth support
- Diverse manufacturer endpoint support
- Audio conference add-on
- Multiple video screen layouts
- Gateway connecting ISDN and IP networks
- Speed matching and true transcoding
- Document conferencing
- Certifications, confirmations, billing, and reporting

The company also offers both hosted and self-service videoconferencing. Managed conferencing includes scheduling, reservations, participant greetings, and technical support. Self-service videoconferencing enables customers to initiate conferences and modify call parameters on demand.

Additional services include public room rentals, satellite connections, conference recording, consulting and training, network provisioning, language capabilities, and dedicated service delivery teams (www.vspan.com).

Backbone Providers

Videoconferencing requires a lot of bandwidth, and the backbone providers are the companies that deliver it. These corporations tend to be large network service providers whose nationwide or even global networks have the bandwidth required to support conferencing applications. Again, this is not a complete list.

The Big Dogs

AT&T, Sprint, and WorldCom dominate the networking world because of their massive global networks that are awash in bandwidth. They provide connectivity via a wide array of protocols, including traditional solutions such as T1-E1, ISDN, and optical, as well as IP. For details about their networks please consult their web sites:

AT&T   www.att.com

Sprint   www.sprint.com

WorldCom   www.worldcom.com

Other Players

There are other players in the game as well who are smaller and more specialized—but no less important. Some of them are described in the following section.

Internap   Internap provides centrally managed Internet connectivity targeted at businesses seeking to maximize the performance of Internet-based applications. The company has created an intelligent routing platform that moves data across the Internet backbone from a single connection to one of the company's service points. By connecting directly to each of the Internet backbone segments, Internap avoids the congested traffic exchange process while using their intelligent routing capability to determine the most direct path across the public infrastructure. Internap began selling Internet connectivity services from its first *Private Network Access Point* (P-NAP) facility in 1996 and has since expanded its network footprint to include Atlanta, Boston, Chicago, Los Angeles, New York, San Jose, and London.

Internap makes TV-quality videoconferencing over the public Internet possible. The company's videoconferencing service is designed to increase enterprise productivity through enhanced and efficient communication, while reducing travel and communication costs. Internap's managed IP connectivity provides the *quality of service* (QoS) necessary for videoconferencing over the public Internet (www.internap.com). Internap's connectivity capabilities are shown below.

Internap Remote Connectivity Services

- T1, fractional DS3, full burstable DS3, OC-3, OC-12
- Route-optimized TCP/IP connectivity through Internap's service points, directly to major Internet backbones
- 24×7 proactive circuit monitoring, outage reporting, and troubleshooting by a dedicated Network Operations Center
- Circuit provisioning
- IP address allocation in compliance with policies of appropriate regional Internet registries

Sonic   Sonic owns and operates a global fiber optic network that is optimized for video transmission. The company offers equipment and transmission services designed for videoconferencing applications and high-bandwidth transmission of multimedia content such as television productions.

The company also facilitates personal and large group videoconferencing (www.sonictelecom.com).

CoreExpress Recently acquired by Williams Communications, CoreExpress offers value-added services that complement those offered by traditional ISPs.

With the CoreExpress Extranet Viewer service management technology, Williams Communications' *virtual private network* (VPN) customers can view and measure network performance. CoreExpress provides backbone transport, network monitoring, and network integration (www.williams.com).

The Service Providers

The service providers are the companies that convert bandwidth and transport sophisticated hardware into a solution that is functionally useful to the user. Bandwidth, after all, is a commodity that can be purchased from dozens of different suppliers who compete for business by offering the lowest price—a sign of a true commodity. Videoconferencing units are sophisticated and feature rich, but they are available from numerous vendors whose prices are equally competitive. Even videoconference bridges are widely available in a variety of forms.

When all of the analysis has been done, the key message that comes out of it is that all customers want to buy is a conference. They don't care about network bandwidth, or about fancy units, or about network service providers. All they care about is that they walk into the room, sit down, and it works. They see the other party or parties on the monitor. In some cases they may not want to own the equipment at all. Instead, they may want to go to a public videoconferencing room and pay someone for the right to use the facilities. For many companies whose videoconferencing use is relatively small, this is an ideal model. Others may want to install a complete conferencing facility in their building, and as long as they use it regularly or can rent it to others, this model works as well. Either way, the service provider segment of the industry provides management and solutions.

ACT Teleconferencing

ACT's suite of applications for audio, video, and Internet conferencing provides easy-to-use communications tools (www.acttel.com). The company's services include

- *ReadyConnect* Audio conference calls without advance reservations or operator assistance.
- *ActionCall* Gives the customer the ability to conduct a high-level audio conference call with full operator support.
- *Passcode* Provides a secured audio conference, based on secure delivery of a passcode.
- *ActionView* The capability for a one-to-many conference via video monitor to participants in worldwide offices.
- *ClarionCall* Delivers full duplex unattended conferencing over your VoIP network.
- *ActionCast* and *ActionCast Plus* Enables a client to stream an audio message over the Internet for an effective conference presentation.
- *ReadyConnect Online* Delivers an "online tour guide," providing the capability to show presentations, guide participants through a web tour, annotate slides in real time, and share applications during a collaborative meeting.

Proximity, Inc.

Recently acquired by ACT Teleconferencing, Proximity services include public access to over 3,500 videoconferencing sites worldwide, large-scale videoconferencing productions, multipoint events, and corporate videoconference consulting and outsourcing. Proximity offers the highest level of personalized customer service, including high-impact videoconferencing events, outsourced corporate videoconferencing functions, and domestic or international videoconferences (www.proximity.com).

Centra

Centra provides web collaboration solutions that enable information delivery in a variety of live and self-service formats and add the capability to record interactions and content, manage information, and reuse it as required across the enterprise.

The company's CentraOne™ web collaboration platform offers real-time web collaboration products, content creation tools, and knowledge delivery systems.

An added benefit is that Centra uses the existing corporate network infrastructure through a simple browser interface. Centra solutions are easily installed using Centra's rapid deployment methodology and easily accessed through Centra's secure, full-service ASP (www.centra.com).

Equant

Equant offers network, integration, and managed services to global business. The company's extensive network connects business centers in 220 countries, with local support in 145 countries. Building on more than 50 years of experience in data communications and as a member of the France Telecom Group, Equant meets the needs of global corporations with an extensive portfolio of managed data network service (www.equant.com).

Glowpoint

Glowpoint is a leading provider of H.323 conferencing capability. The company offers a diversity of services and guarantees, including

- Guaranteed up-time
- Online, real-time billing and usage information
- Gateway services to legacy ISDN-based sites
- Multipoint bridging

- Live operator assistance
- Remote management
- Reduced transmission rates
- International least-cost routing

Working closely with Glowpoint is Wire One Technologies, Inc., a provider of end-to-end videoconferencing solutions. The company's customer base includes more than 3,000 companies with approximately 15,000 videoconferencing endpoints in the commercial, federal and state government, and medical and education marketplaces around the world.

Wire One is a leader in video-over-IP solutions through their relationship with Glowpoint, the nation's first subscriber-based network dedicated to H.323 videoconferencing. The Glowpoint network improves the quality, reliability, and functionality of video communications with lower network costs (www.glowpoint.com).

PlaceWare

PlaceWare's first product, PlaceWare Auditorium, gave customers a live, web-based presentation solution for customer communication scenarios. PlaceWare later became the first company to offer web conferencing as a solution for holding meetings over the Internet.

Today, more than 2,000 companies use PlaceWare for marketing seminars, product launches, training, press tours, customer meetings, and announcements.

PlaceWare requires no additional hardware or software. It relies on nothing more than a phone and web browser, and is extremely easy to use, both as a meeting facilitator and as a participant (www.placeware.com).

WebEx

WebEx provides web-based services that integrate voice, video, and data to create interactive and collaborative applications across a diversity of geographies and platforms. These services are based on the company's multimedia switching platform and are deployed across a global network. WebEx's services permit users to share presentations, documents, voice, and video using Windows, Macintosh, or Solaris systems, all with a web browser (www.webex.com).

PART 3

Conducting a Successful Conference

The Corporate Videoconferencing Facility

As we noted earlier, there are two ways a corporation can have video-conferencing capability. The first is through the use of a videoconferencing service provider that arranges the use of a public room and manages the entire event from start to finish. The other is to purchase and install a videoconferencing room for the corporation. We will discuss each of these options in turn.

The Public Room Option

Companies that use videoconferencing only occasionally cannot justify the cost of installing a dedicated room for the service. Instead, they have the option of contracting with a service provider like Proximity or Kinko's. These companies are true *service* providers: They provide the oversight, scheduling, logistics management, and event execution for the client. If they do their jobs properly, the client walks into the room, sits down, and commences the meeting.

As a customer, there are a number of issues that should be discussed with the service provider. The first of these is fee structure. How is the event to be billed? Is it a fixed price or is it billed on an hourly basis? Does the price include equipment, support, and other ancillary services, or are these separate line items? What are all of the billable components of the service?

The second issue is service quality. Videoconferences today must be high-quality events, or they are deemed to be failures. Does the service provider offer a service level agreement as part of the service that they offer? What quality can you expect from the connection? If possible, have the service provider arranged a demonstration so that you know exactly what to expect. Ask them what kinds of activities do not work well on videoconferences—they are very aware of the kinds of things that work well as well as those that do not. For example, because of the compression that takes place during the transmission of the signal, some information is lost. Although a face is perfectly discernible, a spreadsheet, even when projected on a high-quality document camera, is absolutely unreadable. So take the time to talk to the service provider about recommended use of the technology.

The third issue is availability. What kind of lead-time is required to establish the conference? This will vary with the location and number of sites requested, but give the service provider adequate time to manage the event.

The fourth issue is support. What kinds of services are part of the overall package? For example, will there be technical support readily available in the event of video or audio difficulty? Does the package include added production value such as lighting, backdrops, site support, and materials management? Will there be an on-site coordinator to serve as a single point of contact? If so, what are their responsibilities? (See the site coordinator discussion in Part 1.)

Most videoconferencing service providers are extremely professional and make it their business to manage these details for the client so that the client doesn't have to. However, it doesn't hurt to ask questions and be informed.

Designing and Building a Room

For companies whose videoconferencing usage is consistently high, it makes sense to purchase one or more videoconferencing units and dedicate a room for the purpose. Although it seems like a simple thing to buy a unit, roll it into a room, add a table and a few chairs, and start conferencing, it is in fact significantly more complex than that. Failure to take into consideration additional factors, many of which are less-than-intuitive, can result in a poorly designed and nonfunctional room and a significant waste of capital.

The Room

The single most important factor in the design of a videoconferencing facility is the size of the room. The most common mistake that companies make is installing the equipment in a room that is too small for the videoconferencing application that will take place there. Videoconferencing rooms require a minimum of two to three square feet per participant.

Figure 3-1 shows the layout of a typical videoconference room that is too small. There are eight people in the room, but only four of them are actually visible to the far end because the videoconferencing unit's lens isn't wide enough to capture all of the participants. If it *were* wide enough, the images of the people in the room would be distorted. Of course, some will argue that this can be overcome by panning the camera back and forth, and although this can be done, it is distracting. Furthermore, the seating is not optimal because the participants have to sit at an angle to take part in the conference. Better to pick a larger room as shown in Figure 3-2.

The other alternative is seating. The goal in designing the seating arrangements is to flatten the viewing plane as much as possible to ensure focus across the entire room and to minimize the amount of adjustment that the camera must make as it moves from speakers to speaker.

There are three options for seating participants in a videoconferencing room. The first, shown in Figure 3-3, is to arrange the

Figure 3-1
A videoconference room that is too small

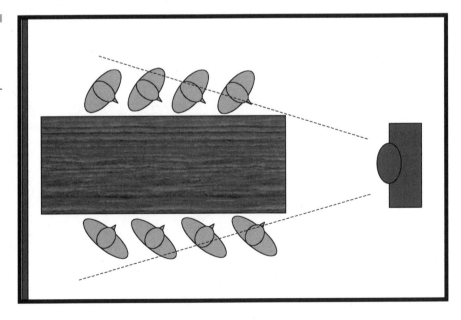

Figure 3-2
A larger videoconference room enables all participants to be seen.

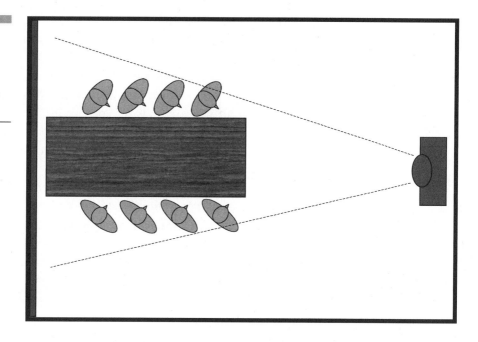

Figure 3-3
Configuration one: a V shape

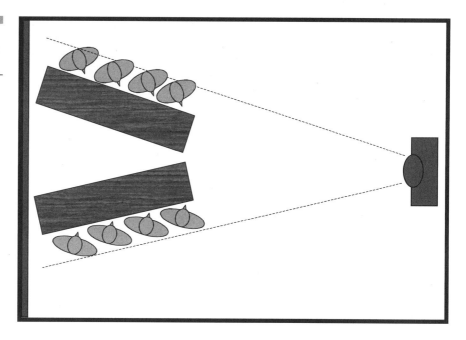

furniture so that the participants sit in a "V" facing the conference unit. This provides good coverage for the camera and comfort for the participants.

The second technique is to seat the participants in a semi-circle, as shown in Figure 3-4. This provides adequate coverage for the camera and again, offers a comfortable seating arrangement for conference participants.

The third technique is to tier the participants, as shown in Figure 3-5.

Another consideration is the actual use to which the room will be put. If the room is to be a dedicated videoconferencing room that will be used for nothing else, that's great. More commonly, however, it will also be used for ad hoc meetings and other purposes, which is fine— but it must be clearly understood that the primary purpose for the room is videoconferencing, and the room design must reflect that.

Many companies feel that the corporate boardroom is the best place to install the videoconferencing room because a lot of money typically went into making it look good. For this very reason board-

Figure 3-4
Second configuration: a semicircle

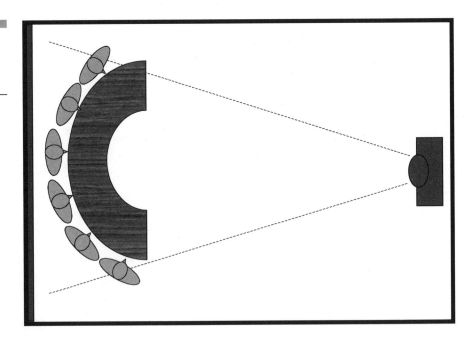

Figure 3-5
Third
configuration:
tiered participants

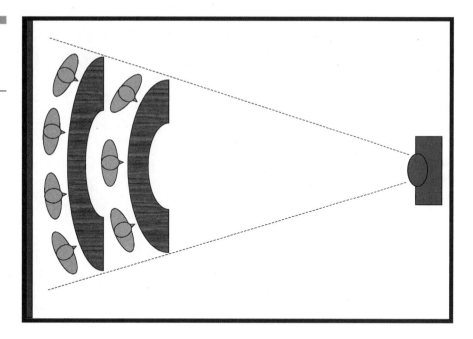

rooms don't usually make good videoconferencing rooms. They tend to be dark and poorly lit, and because a lot of money went into them, the room manager will not be thrilled with the idea of running cables, drilling holes, and changing the lighting and curtains to make the room video-friendly. Better to designate an alternate facility.

Room Color and Lighting

The room should be painted a neutral gray color in flat, nonreflective paint. Although not necessary, some companies hang a royal blue or green curtain behind the head of the table (directly in front of the camera) to add color and a degree of professional "set dressing." Alternatively, a company logo or location identifier (see Figure 3-6) can be added. Even if the location identifier is not on the wall, a sign on the table, as shown in Figure 3-7, should be used to identify participating sites.

Figure 3-6
Curtains and a
corporate logo
identify the
location and
provide
soundproofing.

Figure 3-7
Each location
should have a sign
that uniquely
identifies it.

Room Lighting

The standard fluorescent lights found in most conference rooms cast a decidedly green pall on everyone in the room. For videoconferencing, it is advisable to remove the normal tubes and install bulbs that emit light that emulates the spectrum of natural sunlight. The reflectors in the light fixtures should be replaced so that they bathe the walls and all other surfaces with even lighting. Additional lights, such as those shown in Figure 3-8, can be added to further illuminate the participants. The angle of the lights should be adjustable, and all lights should be on dimmers to enable the ambient lighting to be adjusted with significant granularity.

Room Acoustics

In most cases, the room that will be converted for videoconferencing will not be soundproof. Of course, the room is not designed to be a sound stage; as long as ambient noise from the surroundings is

Figure 3-8
Both fixed and adjustable lights can be used to illuminate the room.

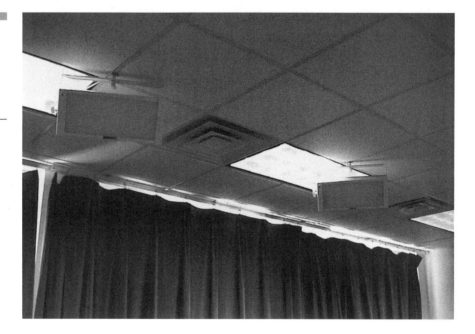

reduced, the room will suffice. However, the room should be chosen carefully: designers should consider baseline noise from air handlers, fluorescent fixtures, and adjacent noisy rooms (lunchrooms, lounges), in addition to street noise.

Noise can be reduced in a number of ways. Hanging heavy curtains around the room, for example, can dramatically reduce the amount of sound that enters from the outside. The curtains fulfill two other functions as well: they look good, and they block light coming in from outside if there are windows in the room.

The room should be covered with a neutral color, short nap carpet to further reduce sound bounce and to eliminate the noise of shoes clicking across a hard floor.

Furniture

Furniture should be chosen with several criteria in mind. It should be neutral in color, preferably light gray without discernible patterns, and should have nonreflective surfaces. It should be easily movable to facilitate rearrangement for different conference scenarios.

Equipment

As Tim Allen would say, "You can't have enough tools." That's certainly true in the videoconference room.

First and foremost is the videoconference unit, an example of which is the Tandberg unit shown in Figure 3-9. This particular unit has two monitors so that it can simultaneously display both the near and far-end locations of a conference. The second monitor is optional but is highly recommended. In every case, the unit should be easily upgradeable (CODEC software). A useful add-on for the videoconference unit is a cart; being able to move the device easily facilitates rearrangement of the room when necessary.

Other equipment is also optional but recommended. A document camera, such as the one shown in Figure 3-10, enables a participant to draw on a piece of paper and have the camera transmit the video

Figure 3-9
Videoconferencing units like this one can be large; allow enough space for it and other equipment.

Figure 3-10
A document camera. Its output can be transmitted over the video network.

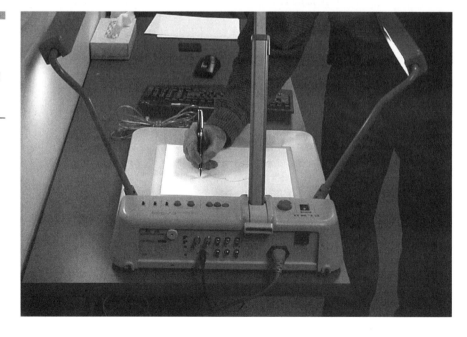

image into the system so that all participants can see it. These devices work extremely well for displaying objects or for displaying simple line drawings. They do not work well for printed text or for spreadsheets because of the loss of image detail due to signal compression.

Another tool that is extremely useful in a videoconferencing facility is a scan converter, which converts a signal from a PC so that it can be displayed on the TV screen of a videoconferencing unit.

Finally, some kind of video playback unit is useful—a VCR or DVD, for example. A VCR enables the conference to be taped, but be careful: Wiretap laws in many countries require that all participants be clearly advised that they are being recorded. Do not fail to make an announcement if you intend to record the session.

Network Connections

The bulk of videoconferencing units today use ISDN for connectivity with the transport network. Room designers should check with their IT staff or with their local service provider to ensure that the appropriate network connection will be available.

As we noted earlier in the book, be wary of commingling video traffic with other corporate data streams. Video is a bandwidth hog and will create problems on the corporate LAN if it is enabled to travel there. Best to segregate the two to ensure maximum performance. A single videoconference will consume 840Kbps of bandwidth; on a bursty LAN, that can seriously affect service.

One area that is often discussed is the question of encryption. For the most part, video signals travel across the public switched telephone network, where they could conceivably be intercepted in the same way a telephone call could be monitored. In order to do that, the would-be interloper would have to either be in the equipment closet in the building where the conference is taking place or in the telephone company central office. The point is that although the possibility of signal interception exists, the chances are so remote that it is probably not worth worrying about.

Of course, encryption is available, but in order for it to work, all sites must have the ability to both encrypt and decrypt the signal. This capability adds significant expense to each location.

Other Considerations

Videoconferencing equipment represents a significant capital investment. As a result, the room should be kept locked at all times. The room should have a single person responsible for its use to eliminate scheduling conflicts and to ensure that the room is prepared at all times for use. Locking the room also prevents unauthorized people from using the equipment. It isn't a television, even though it may look like one. Bob Maurer of Proximity recounts an endless chain of stories about trouble calls that he has fielded, all the result of people monkeying around with the equipment.

The Conference

To ensure a successful conference, attendees should follow a simple set of guidelines that are not all that different from those listed earlier in the book for television studio presenters.

Participants should wear solid colors, preferably blues, grays, and greens, *never* red, and should avoid clothing with patterns. They should plan to arrive a bit early to familiarize themselves with the environment and to ask last-minute questions about the equipment.

As with any meeting, participants should come prepared for the task at hand. Remember, it's just a meeting—the distance component should be relatively invisible if the conference is established and conducted properly.

The room coordinator should take a few minutes to familiarize participants with the use of the equipment, the features of the system, and who they should call in the event of a problem.

If these guidelines and those listed earlier for the studio presenter are followed, the conference will be successful.

IN CONCLUSION

So what have we learned from all this? First of all, there is no question that although the extraordinary usage growth of conferencing services that occurred in the days following September 11th has settled back down, usage is still significantly higher than it was before the event—and does not show any sign of abating. Because of available bandwidth, well-designed applications, and a strong desire on the part of customers to use alternative delivery solutions, video and other related conferencing technologies have come into their own.

Four factors drive this desire. The first is a growing need for instantaneous, effortless, and universal access to information. In today's "Knowledge Age," competitive advantage is determined by the speed and accuracy with which a company can access its corporate databases and other sources of tactical and strategic information. The company that gets there first wins the game.

The second requirement is rich content. No longer is it adequate to provide information in the stark black and white of formatted text. System users want text, but they also want graphics, full color photographs, and video clips. People respond faster and more accurately to information that tickles more than a single sense; the message of multimedia, therefore, can be extremely powerful.

The third requirement is cost-effective, innovative delivery of information. Customers are willing to pay more for multimedia information, but the additional cost must be justified by an efficient, effective interface that *clearly* impacts the business in a positive fashion.

The final requirement, of course, is a system interface that is easy to use and that provides instantaneous, painless access to information. If these four requirements can be met, the effort will succeed. However, it requires the combined efforts of all four layers of the marketplace—the system manufacturers, the bridge providers, the backbone providers, and the service providers—with particular emphasis on the service providers—to complete the customer service equation. Some of these segments will become commodities to one degree or another because commodities are only differentiable on price. To be successful, to be uniquely differentiated, requires that a company offer something that others do not. In the videoconferencing business, that means offering a set of converged services that make the technology transparent to the user. That is a tough challenge but one that service providers must take on if they are to drive videoconferencing to critical mass.

APPENDIX A

A Brief History of Videoconferencing

Videoconferencing has a colorful history that is worth examining. First introduced in the 1960s to enthusiastic crowds who saw technologies introduced on the Jetsons to be on the verge of becoming a reality, it disappeared about as quickly as it arrived. For all the pomp, circumstance, and media hype that accompanied its introduction, videoconferencing initially failed. In fact, AT&T spent more than a billion dollars on the concept before throwing in the towel.

Hugo Gernsback was born in 1884 in Luxembourg City, Luxembourg. He emigrated to America in 1904 and became a U.S. citizen. His magazine, *Amazing Stories*, began publication in 1926 as the first periodical devoted to fiction of the future and was followed by *Wonder Stories* in 1929. The prestigious Hugo Award for science fiction excellence was named for him, as was a crater on the Moon.

Gernsback also published a magazine called *Modern Electrics*. In 1911, he published a serial story in *Modern Electrics* called "Ralph 124C 411" in which he regaled readers with a technologically rich future, including the first reference to an application that would ultimately become videoconferencing. The story opens as one of the world's great scientists, Ralph, is talking with a friend from New York using a device called the *telephot*. Suddenly, the call is interrupted, and the telephot screen goes blank. When the image returns, his friend's face has been replaced with that of a beautiful woman clad in evening attire. She is clearly shocked by the face of the strange man gazing upon her. Ralph responds, "I beg your pardon, but 'Central' seems to have made another mistake. I shall certainly have to make a complaint about the service."

The first *real* videoconferencing system to gain the attention of the world was AT&T's PicturePhone. With AT&T's limitless financial and technological resources, the system was well conceived and designed and was based on leading-edge technologies that were just emerging.

The first system, built in 1956, was rather primitive. It transmitted one frame (image) every two seconds (compared to modern video systems that transmit 30 frames per second, or 60 times the rate of the first PicturePhone). By 1964, an experimental commercial system, the "Mod 1," had been developed at Bell Laboratories. It was installed at Disneyland and the New York World's Fair, where attendees were invited to place calls between the two exhibits. After placing calls, visitors were interviewed by a market research agency to gauge the nature of their experience.

The product had everything going for it and looked to be heading toward success. The buying public, enthralled with the heady technologies that filled their world at the time (the space program, most notably), saw the video phone as the realization of that which television told them was truly possible. There were even tangible uses of the PicturePhone: Hugh Hefner installed them at the Playboy mansion so that he could review photos of potential Playmates from various remote locations.

Of course, the law of unintended consequences soon reared its ugly head. Even though *very* few people were able to buy them, excitement about PicturePhone's capabilities was rampant. At the World's Fair, people lined up for hours to make PicturePhone calls. Even First Lady Ladybird Johnson tried the devices.

In 1970, following years of market research, AT&T rolled out the "Mod II." The service was first introduced in downtown Pittsburgh, later in Washington and Chicago, and AT&T executives predicted that they would sell a million systems by 1980.

To their chagrin, PicturePhone received an underwhelming response. The reasons were all too clear: ridiculously high subscription rates ($125 a month in the 1970s), the equipment was far too bulky, the controls were not user-friendly, and the picture was far too small (image size was 4.5 by 5.75 inches). Fears of privacy, such as those illustrated by the wrong number in Gernsback's novel, began to creep into the public psyche, but one issue overwhelmed all others: because very few people could afford to own a PicturePhone, what were the chances that there would be someone else to talk to who also owned a system?

It would take another 25 years before person-to-person videoconferencing would actually catch on. AT&T's PicturePhone was clearly ahead of its time, and although the service was available in numerous cities, it never became popular. In addition to the sociological and economic reasons, there was a sound technological basis for this as well: keep in mind that these systems were introduced long before the arrival of PCs, microprocessors, video CODECs, or data compression. Digital transmission had arrived in 1962 with the introduction of T-Carrier, but digital was truly in its infancy.

The PicturePhone also required substantial physical network resources to work properly. It was connected to the Central Office via three wire pairs instead of the single pair required by traditional telephony. One pair carried the 1 MHz video signal in one direction, while the other carried the video in the opposite direction. These pairs had to be conditioned to carry the video signals. The third pair carried the voice signal.

Central Offices equipped for PicturePhone service had a separate electromechanical crossbar switch reserved for video calls. When a call had to be connected to someone served via another Central Office, things became remarkably complex. Telephone calls are typically multiplexed across a carrier system when transmitted from one office to another. The carrier can be microwave, coaxial cable, or fiber. A voice channel requires 4 KHz for satisfactory transmission, while a PicturePhone call requires more than 1 MHz. (250 times the bandwidth!) Clearly, a small number of video calls would rapidly exhaust all available bandwidth, a great example of the dictum, "In its success lie the seeds of its own destruction."

Shortly after the arrival of the microprocessor and the PC, videoconferencing began to emerge as a viable application. The semiconductor made it possible to build video CODECs for reasonable prices, and soon a number of vendors began to sell videoconferencing units. These included Intel, PictureTel, and Polycom, to name a few. Today the technology is fairly universal and is widely used for a variety of applications.

APPENDIX B

Common Industry Acronyms

AAL	ATM Adaptation Layer
AARP	AppleTalk Address Resolution Protocol
ABM	Asynchronous Balanced Mode
ABR	Available bit rate
AC	Alternating Current
ACD	Automatic Call Distribution
ACELP	Algebraic Code-Excited Linear Prediction
ACF	Advanced Communication Function
ACK	Acknowledgment
ACM	Address Complete Message
ACSE	Association Control Service Element
ACTLU	Activate Logical Unit
ACTPU	Activate Physical Unit
ADCCP	Advanced Data Communications Control Procedures
ADM	Add/Drop Multiplexer
ADPCM	Adaptive Differential Pulse Code Modulation
ADSL	Asymmetric Digital Subscriber Line
AFI	Authority and Format Identifier
AIN	Advanced Intelligent Network
AIS	Alarm Indication Signal
ALU	Arithmetic-logic unit
AM	Administrative Module (Lucent 5ESS)
AM	Amplitude Modulation
AMI	Alternate Mark Inversion
AMP	Administrative Module Processor
AMPS	Advanced Mobile Phone System
ANI	Automatic Number Identification (SS7)
ANSI	American National Standards Institute
APD	Avalanche Photodiode

API	Application Programming Interface
APPC	Advanced Program-to-Program Communication
APPN	Advanced Peer-to-Peer Networking
APS	Automatic Protection Switching
ARE	All Routes Explorer (Source Route Bridging)
ARM	Asynchronous Response Mode
ARP	Address Resolution Protocol (IETF)
ARPA	Advanced Research Projects Agency
ARPANET	Advanced Research Projects Agency Network
ARQ	Automatic Repeat Request
ASCII	American Standard Code for Information Interchange
ASI	Alternate Space Inversion
ASIC	Application Specific Integrated Circuit
ASK	Amplitude Shift Keying
ASN	Abstract Syntax Notation
ASP	Application Service Provider
AT&T	American Telephone and Telegraph
ATDM	Asynchronous Time Division Multiplexing
ATM	Asynchronous Transfer Mode
ATM	Automatic Teller Machine
ATMF	Asynchronous Transfer Mode Forum
AU	Administrative Unit (SDH)
AUG	Administrative Unit Group (SDH)
AWG	American Wire Gauge
B8ZS	Binary 8 Zero Substitution
BANCS	Bell Administrative Network Communications System
BBN	Bolt, Beranak, and Newman
BBS	Bulletin Board Service
Bc	Committed Burst Size
BCC	Blocked Calls Cleared
BCC	Block Check Character
BCD	Blocked Calls Delayed
BCDIC	Binary Coded Decimal Interchange Code

Be	Excess Burst Size
BECN	Backward Explicit Congestion Notification
BER	Bit Error Rate
BERT	Bit Error Rate Test
BGP	Border Gateway Protocol (IETF)
BIB	Backward Indicator Bit (SS7)
B-ICI	Broadband Intercarrier Interface
BIOS	Basic Input/Output System
BIP	Bit Interleaved Parity
B-ISDN	Broadband Integrated Services Digital Network
BISYNC	Binary Synchronous Communications Protocol
BITNET	Because It's Time Network
BITS	Building Integrated Timing Supply
BLSR	Bidirectional Line Switched Ring
BOC	Bell Operating Company
BPRZ	Bipolar Return to Zero
Bps	Bits per second
BRI	Basic Rate Interface
BRITE	Basic Rate Interface Transmission Equipment
BSC	Binary Synchronous Communications
BSN	Backward Sequence Number (SS7)
BSRF	Bell System Reference Frequency
BTAM	Basic Telecommunications Access Method
BUS	Broadcast Unknown Server
C/R	Command/Response
CAD	Computer-Aided Design
CAE	Computer-Aided Engineering
CAM	Computer-Aided Manufacturing
CAP	Carrierless Amplitude/Phase Modulation
CAP	Competitive Access Provider
CARICOM	Caribbean Community and Common Market
CASE	Common Application Service Element
CASE	Computer-Aided Software Engineering

CAT	Computer-Aided Tomography
CATIA	Computer-Assisted Three-dimensional Interactive Application
CATV	Community Antenna Television
CBEMA	Computer and Business Equipment Manufacturers Association
CBR	Constant Bit Rate
CBT	Computer-based training
CC	Cluster Controller
CCIR	International Radio Consultative Committee
CCIS	Common Channel Interoffice Signaling
CCITT	International Telegraph and Telephone Consultative Committee
CCS	Common Channel Signaling
CCS	Hundred Call Seconds per Hour
CD	Collision Detection
CD	Compact Disc
CDC	Control Data Corporation
CDMA	Code Division Multiple Access
CDPD	Cellular Digital Packet Data
CD-ROM	Compact Disc-Read Only Memory
CDVT	Cell delay variation tolerance
CEI	Comparably Efficient Interconnection
CEPT	Conference of European Postal and Telecommunications Administrations
CERN	European Council for Nuclear Research
CERT	Computer Emergency Response Team
CES	Circuit emulation service
CEV	Controlled Environmental Vault
CGI	Common Gateway Interface (Internet)
CHAP	Challenge Handshake Authentication Protocol
CICS	Customer Information Control System
CICS/VS	Customer Information Control System/Virtual Storage
CIDR	Classless Interdomain Routing (IETF)

CIF	Cells in Frames
CIR	Committed Information Rate
CISC	Complex Instruction Set Computer
CIX	Commercial Internet Exchange
CLASS	Custom Local Area Signaling Services (Bellcore)
CLEC	Competitive Local Exchange Carrier
CLLM	Consolidated Link Layer Management
CLNP	Connectionless Network Protocol
CLNS	Connectionless Network Service
CLP	Cell Loss Priority
CM	Communications Module (Lucent 5ESS)
CMIP	Common Management Information Protocol
CMISE	Common Management Information Service Element
CMOL	CMIP over LLC
CMOS	Complementary Metal Oxide Semiconductor
CMOT	CMIP over TCP/IP
CMP	Communications Module Processor
CNE	Certified NetWare Engineer
CNM	Customer Network Management
CNR	Carrier-to-Noise Ratio
CO	Central Office
CoCOM	Coordinating Committee on Export Controls
CODEC	Coder/decoder
COMC	Communications Controller
CONS	Connection-Oriented Network Service
CORBA	Common Object Request Brokered Architecture
COS	Class of Service (APPN)
COS	Corporation for Open Systems
CPE	Customer premises equipment
CPU	Central processing unit
CRC	Cyclic Redundancy Check
CRT	Cathode Ray Tube
CRV	Call Reference Value

CS	Convergence Sublayer
CSA	Carrier Serving Area
CSMA	Carrier Sense Multiple Access
CSMA/CA	Carrier Sense Multiple Access with Collision Avoidance
CSMA/CD	Carrier Sense Multiple Access with Collision Detection
CSU	Channel Service Unit
CTI	Computer Telephony Integration
CTIA	Cellular Telecommunications Industry Association
CTS	Clear to Send
CU	Control Unit
CVSD	Continuously Variable Slope Delta Modulation
CWDM	Coarse Wavelength Division Multiplexing
D/A	Digital-to-Analog
DA	Destination Address
DAC	Dual Attachment Concentrator (FDDI)
DACS	Digital Access and Cross-connect System
DARPA	Defense Advanced Research Projects Agency
DAS	Dual Attachment Station (FDDI)
DASD	Direct Access Storage Device
dB	Decibel
DBS	Direct Broadcast Satellite
DC	Direct Current
DCC	Data Communications Channel (SONET)
DCE	Data Circuit-terminating Equipment
DCN	Data Communications Network
DCS	Digital Cross-connect System
DCT	Discrete Cosine Transform
DDCMP	Digital Data Communications Management Protocol (DNA)
DDD	Direct Distance Dialing
DDP	Datagram Delivery Protocol
DDS	DATAPHONE Digital Service (Sometimes Digital Data Service)
DE	Discard eligibility (LAPF)

DECT	Digital European Cordless Telephone
DES	Data Encryption Standard (NIST)
DID	Direct Inward Dialing
DIP	Dual Inline Package
DLC	Digital Loop Carrier
DLCI	Data Link Connection Identifier
DLE	Data Link Escape
DLSw	Data Link Switching
DM	Delta Modulation
DM	Disconnected Mode
DMA	Direct Memory Access (computers)
DMAC	Direct Memory Access Control
DME	Distributed Management Environment
DMS	Digital Multiplex Switch
DNA	Digital Network Architecture
DNIC	Data Network Identification Code (X.121)
DNIS	Dialed Number Identification Service
DNS	Domain Name System (IETF)
DOD	Direct Outward Dialing
DOD	Department of Defense
DOJ	Department of Justice
DOV	Data over Voice
DPSK	Differential Phase Shift Keying
DQDB	Distributed Queue Dual Bus
DRAM	Dynamic Random Access Memory
DSAP	Destination Service Access Point
DSF	Dispersion-Shifted Fiber
DSI	Digital Speech Interpolation
DSL	Digital Subscriber Line
DSLAM	Digital Subscriber Line Access Multiplexer
DSP	Digital Signal Processing
DSR	Data Set Ready
DSS	Digital Satellite System

DSS	Digital Subscriber Signaling System
DSU	Data Service Unit
DTE	Data Terminal Equipment
DTMF	Dual Tone Multifrequency
DTR	Data Terminal Ready
DV	Digital Video
DV	Distance Vector
DVRN	Dense Virtual Routed Networking (Crescent)
DWDM	Dense Wavelength Division Multiplexing
DXI	Data Exchange Interface
E/O	Electrical-to-Optical
EBCDIC	Extended Binary Coded Decimal Interchange Code
ECMA	European Computer Manufacturer Association
ECN	Explicit Congestion Notification
ECSA	Exchange Carriers Standards Association
EDFA	Erbium-Doped Fiber Amplifier
EDI	Electronic Data Interchange
EDIBANX	EDI Bank Alliance Network Exchange
EDIFACT	Electronic Data Interchange for Administration, Commerce, and Trade (ANSI)
EFCI	Explicit Forward Congestion Indicator
EFTA	European Free Trade Association
EGP	Exterior Gateway Protocol (IETF)
EIA	Electronics Industry Association
EIGRP	Enhanced Interior Gateway Routing Protocol
EIR	Excess Information Rate
EMBARC	Electronic Mail Broadcast to a Roaming Computer
EMI	Electromagnetic Interference
EMS	Element Management System
EN	End Node
ENIAC	Electronic Numerical Integrator and Computer
EO	End Office
EOC	Embedded Operations Channel (SONET)

EOT	End of Transmission (BISYNC)
EPROM	Erasable Programmable Read Only Memory
ESCON	Enterprise System Connection (IBM)
ESF	Extended Superframe Format
ESP	Enhanced Service Provider
ESS	Electronic Switching System
ETSI	European Telecommunications Standards Institute
ETX	End of Text (BISYNC)
EWOS	European Workshop for Open Systems
FACTR	Fujitsu Access and Transport System
FAQ	Frequently Asked Question
FAT	File Allocation Table
FCS	Frame Check Sequence
FDD	Frequency Division Duplex
FDDI	Fiber Distributed Data Interface
FDM	Frequency Division Multiplexing
FDMA	Frequency Division Multiple Access
FDX	Full-Duplex
FEBE	Far End Block Error (SONET)
FEC	Forward Error Correction
FEC	Forward Equivalence Class
FECN	Forward Explicit Congestion Notification
FEP	Front-End Processor
FERF	Far End Receive Failure (SONET)
FET	Field Effect Transistor
FHSS	Frequency hopping spread spectrum
FIB	Forward Indicator Bit (SS7)
FIFO	First in First out
FITL	Fiber in the Loop
FLAG	Fiber Ling Across the Globe
FM	Frequency Modulation
FPGA	Field Programmable Gate Array
FR	Frame Relay

FRAD	Frame relay access device
FRBS	Frame Relay Bearer Service
FSK	Frequency Shift Keying
FSN	Forward Sequence Number (SS7)
FTAM	File Transfer, Access, and Management
FTP	File Transfer Protocol (IETF)
FTTC	Fiber to the Curb
FTTH	Fiber to the Home
FUNI	Frame User-to-Network Interface
FWM	Four Wave Mixing
GATT	General Agreement on Tariffs and Trade
GbE	Gigabit Ethernet
Gbps	Gigabits per second (billions of bits per second)
GDMO	Guidelines for the Development of Managed Objects
GEOS	Geosynchronous Earth Orbit Satellites
GFC	Generic Flow Control (ATM)
GFI	General Format Identifier (X.25)
GOSIP	Government Open Systems Interconnection Profile
GPS	Global Positioning System
GRIN	Graded Index (fiber)
GSM	Global System for Mobile Communications
GUI	Graphical user interface
HDB3	High Density, Bipolar 3 (E-Carrier)
HDLC	High-level Data Link Control
HDSL	High-bit-rate Digital Subscriber Line
HDTV	High Definition Television
HDX	Half-Duplex
HEC	Header Error Control (ATM)
HFC	Hybrid Fiber/Coax
HFS	Hierarchical File Storage
HLR	Home Location Register
HSSI	High-Speed Serial Interface (ANSI)
HTML	Hypertext Markup Language

HTTP	Hypertext Transfer Protocol (IETF)
HTU	HDSL Transmission Unit
I	Intrapictures
IAB	Internet Architecture Board (formerly Internet Activities Board)
IACS	Integrated Access and Cross-connect System
IAD	Integrated Access Device
IAM	Initial Address Message (SS7)
IANA	Internet Address Naming Authority
ICMP	Internet Control Message Protocol (IETF)
IDP	Internet Datagram Protocol
IEC	Interexchange Carrier (also IXC)
IEC	International Electrotechnical Commission
IEEE	Institute of Electrical and Electronics Engineers
IETF	Internet Engineering Task Force
IFRB	International Frequency Registration Board
IGP	Interior Gateway Protocol (IETF)
IGRP	Interior Gateway Routing Protocol
ILEC	Incumbent Local Exchange Carrier
IML	Initial Microcode Load
IMP	Interface Message Processor (ARPANET)
IMS	Information Management System
InARP	Inverse Address Resolution Protocol (IETF)
InATMARP	Inverse ATM Address Resolution Protocol
INMARSAT	International Maritime Satellite Organization
INP	Internet Nodal Processor
InterNIC	Internet Network Information Center
IP	Internet Protocol (IETF)
IPX	Internetwork Packet Exchange (NetWare)
ISDN	Integrated Services Digital Network
ISO	International Organization for Standardization
ISOC	Internet Society
ISP	Internet Service Provider

ISUP	ISDN User Part (SS7)
IT	Information Technology
ITU	International Telecommunication Union
ITU-R	International Telecommunication Union-Radio Communication Sector
IVD	Inside Vapor Deposition
IVR	Interactive Voice Response
IXC	Interexchange Carrier (also IEC)
JEPI	Joint Electronic Paynets Initiative
JES	Job Entry System
JIT	Just in Time
JPEG	Joint Photographic Experts Group
KB	Kilobytes
Kbps	Kilobits per second (thousands of bits per second)
KLTN	Potassium Lithium Tantalate Niobate
LAN	Local Area Network
LANE	LAN emulation
LAP	Link Access Procedure (X.25)
LAPB	Link Access Procedure Balanced (X.25)
LAPD	Link Access Procedure for the D-Channel
LAPF	Link Access Procedure to Frame Mode Bearer Services
LAPF-Core	Core Aspects of the Link Access Procedure to Frame Mode Bearer Services
LAPM	Link Access Procedure for Modems
LAPX	Link Access Procedure Half-Duplex
LASER	Light Amplification by the Stimulated Emission of Radiation
LATA	Local Access and Transport Area
LCD	Liquid Crystal Display
LCGN	Logical Channel Group Number
LCM	Line Concentrator Module
LCN	Local Communications Network
LD	Laser Diode

LDAP	Lightweight Directory Access Protocol (X.500)
LEAF®	Large Effective Area Fiber® (Corning product)
LEC	Local Exchange Carrier
LED	Light-emitting diode
LENS	Lightwave Efficient Network Solution (Centerpoint)
LEOS	Low Earth Orbit Satellites
LI	Length Indicator
LIDB	Line Information Database
LIFO	Last in First out
LIS	Logical IP Subnet
LLC	Logical Link Control
LMDS	Local Multipoint Distribution System
LMI	Local Management Interface
LMOS	Loop Maintenance Operations System
LORAN	Long-range Radio Navigation
LPC	Linear Predictive Coding
LPP	Lightweight Presentation Protocol
LRC	Longitudinal Redundancy Check (BISYNC)
LS	Link State
LSI	Large Scale Integration
LSP	Label-sSwitched path
LU	Line Unit
LU	Logical Unit (SNA)
MAC	Media Access Control
MAN	Metropolitan Area Network
MAP	Manufacturing Automation Protocol
MAU	Medium Attachment Unit (Ethernet)
MAU	Multistation Access Unit (Token Ring)
MB	Megabytes
MBA™	Metro Business Access™ (Ocular)
Mbps	Megabits per second (millions of bits per second)
MD	Message Digest (MD2, MD4, MD5) (IETF)
MDF	Main Distribution Frame

MEMS	Micro Electrical Mechanical System
MF	Multifrequency
MFJ	Modified Final Judgment
MHS	Message Handling System (X.400)
MIB	Management Information Base
MIC	Medium Interface Connector (FDDI)
MIME	Multipurpose Internet Mail Extensions (IETF)
MIPS	Millions of Instructions Per Second
MIS	Management Information Systems
MITI	Ministry of International Trade and Industry (Japan)
ML-PPP	Multilink Point-to-Point Protocol
MMDS	Multichannel, Multipoint Distribution System
MMF	Multimode Fiber
MNP	Microcom Networking Protocol
MP	Multilink PPP
MPEG	Motion Picture Experts Group
MPLS	Multiprotocol Label Switching
MPOA	Multiprotocol over ATM
MRI	Magnetic Resonance Imaging
MSB	Most Significant Bit
MSC	Mobile Switching Center
MSO	Mobile Switching Office
MSVC	Meta-Signaling Virtual Channel
MTA	Major Trading Area
MTBF	Mean Time Between Failure
MTP	Message Transfer Part (SS7)
MTTR	Mean Time to Repair
MTU	Maximum Transmission Unit
MVS	Multiple Virtual Storage
NAFTA	North American Free Trade Agreement
NAK	Negative Acknowledgment (BISYNC, DDCMP)
NAP	Network Access Point (Internet)
NARUC	National Association of Regulatory Utility Commissioners

NASA	National Aeronautics and Space Administration
NASDAQ	National Association of Securities Dealers Automated Quotations
NATA	North American Telecommunications Association
NATO	North Atlantic Treaty Organization
NAU	Network Accessible Unit
NCP	Network Control Program
NCSA	National Center for Supercomputer Applications
NCTA	National Cable Television Association
NDIS	Network Driver Interface Specifications
NDSF	Non-Dispersion-Shifted Fiber
NetBEUI	NetBIOS Extended User Interface
NetBIOS	Network Basic Input/Output System
NFS	Network File System (Sun)
NIC	Network interface card
NII	National Information Infrastructure
NIST	National Institute of Standards and Technology (formerly NBS)
NIU	Network Interface Unit
NLPID	Network Layer Protocol Identifier
NLSP	NetWare Link Services Protocol
NM	Network Module
Nm	Nanometer
NMC	Network Management Center
NMS	Network Management System
NMT	Nordic Mobile Telephone
NMVT	Network Management Vector Transport Protocol
NNI	Network Node Interface
NNI	Network-to-Network Interface
NOC	Network operations center
NOCC	Network Operations Control Center
NOS	Network Operating System
NPA	Numbering Plan Area

NREN	National Research and Education Network
NRZ	Non-Return to Zero
NRZI	Non-Return to Zero Inverted
NSA	National Security Agency
NSAP	Network Service Access Point
NSAPA	Network Service Access Point Address
NSF	National Science Foundation
NTSC	National Television Systems Committee
NTT	Nippon Telephone and Telegraph
NVOD	Near Video on Demand
NZDSF	Non-Zero Dispersion-Shifted Fiber
OADM	Optical Add-Drop Multiplexer
OAM	Operations, Administration, and Maintenance
OAM&P	Operations, Administration, Maintenance, and Provisioning
OAN	Optical Area Network
OC	Optical Carrier
OEM	Original Equipment Manufacturer
O-E-O	Optical-Electrical-Optical
OLS	Optical Line System (Lucent)
OMAP	Operations, Maintenance, and Administration Part (SS7)
ONA	Open Network Architecture
ONU	Optical Network Unit
OOF	Out of Frame
OS	Operating System
OSF	Open Software Foundation
OSI	Open Systems Interconnection (ISO, ITU-T)
OSI RM	Open Systems Interconnection Reference Model
OSPF	Open Shortest Path First (IETF)
OSS	Operation support systems
OTDM	Optical Time Division Multiplexing
OTDR	Optical Time-Domain Reflectometer
OUI	Organizationally Unique Identifier (SNAP)

OVD	Outside Vapor Deposition
P/F	Poll/Final (HDLC)
PAD	Packet Assembler/Disassembler (X.25)
PAL	Phase Alternate Line
PAM	Pulse Amplitude Modulation
PANS	Pretty Amazing New Stuff
PBX	Private branch exchange
PCM	Pulse Code Modulation
PCMCIA	Personal Computer Memory Card International Association
PCN	Personal Communications Network
PCS	Personal Communications Service
PDA	Personal digital assistant
PDH	Plesiochronous Digital Hierarchy
PDU	Protocol Data Unit
PIN	Positive-Intrinsic-Negative
PING	Packet Internet Groper (TCP/IP)
PLCP	Physical Layer Convergence Protocol
PLP	Packet Layer Protocol (X.25)
PM	Phase Modulation
PMD	Physical Medium Dependent (FDDI)
PNNI	Private Network Node Interface (ATM)
PON	Passive Optical Networking
POP	Point of Presence
POSIT	Profiles for Open Systems Interworking Technologies
POSIX	Portable Operating System Interface for UNIX
POTS	Plain Old Telephone Service
PPP	Point-to-Point Protocol (IETF)
PRC	Primary Reference Clock
PRI	Primary Rate Interface
PROFS	Professional Office System
PROM	Programmable Read-Only Memory
PSDN	Packet Switched Data Network

PSK	Phase Shift Keying
PSPDN	Packet Switched Public Data Network
PSTN	Public Switched Telephone Network
PTI	Payload Type Identifier (ATM)
PTT	Post, Telephone, and Telegraph
PU	Physical Unit (SNA)
PUC	Public Utility Commission
PVC	Permanent virtual circuit
QAM	Quadrature Amplitude Modulation
Q-bit	Qualified data bit (X.25)
QLLC	Qualified Logical Link Control (SNA)
QoS	Quality of service
QPSK	Quadrature Phase Shift Keying
QPSX	Queued Packet Synchronous Exchange
R&D	Research & Development
RADSL	Rate Adaptive Digital Subscriber Line
RAID	Redundant Array of Inexpensive Disks
RAM	Random access memory
RARP	Reverse Address Resolution Protocol (IETF)
RAS	Remote Access Server
RBOC	Regional Bell Operating Company
RF	Radio Frequency
RFC	Request for Comments (IETF)
RFH	Remote Frame Handler (ISDN)
RFI	Radio Frequency Interference
RFP	Request for Proposal
RGB	Red, Green, Blue
RHC	Regional Holding Company
RHK	Ryan, Hankin, and Kent (Consultancy)
RIP	Routing Information Protocol (IETF)
RISC	Reduced Instruction Set Computer
RJE	Remote Job Entry
RNR	Receive Not Ready (HDLC)

ROM	Read-Only Memory
ROSE	Remote Operation Service Element
RPC	Remote Procedure Call
RR	Receive Ready (HDLC)
RTS	Request to Send (EIA-232-E)
S/DMS	SONET/Digital Multiplex System
S/N	Signal-to-Noise Ratio
SAA	Systems Application Architecture (IBM)
SAAL	Signaling ATM Adaptation Layer (ATM)
SABM	Set Asynchronous Balanced Mode (HDLC)
SABME	Set Asynchronous Balanced Mode Extended (HDLC)
SAC	Single Attachment Concentrator (FDDI)
SAN	Storage Area Network
SAP	Service Access Point (generic)
SAPI	Service Access Point Identifier (LAPD)
SAR	Segmentation and Reassembly (ATM)
SAS	Single Attachment Station (FDDI)
SASE	Specific Applications Service Element (subset of CASE, Application Layer)
SATAN	System Administrator Tool for Analyzing Networks
SBS	Stimulated Brillouin Scattering
SCCP	Signaling Connection Control Point (SS7)
SCP	Service Control Point (SS7)
SCREAM™	Scalable Control of a Rearrangeable Extensible Array of Mirrors™ (Calient)
SCSI	Small Computer Systems Interface
SCTE	Serial Clock Transmit External (EIA-232-E)
SDH	Synchronous Digital Hierarchy (ITU-T)
SDLC	Synchronous Data Link Control (IBM)
SDS	Scientific Data Systems
SECAM	Sequential Color with Memory
SF	Superframe Format (T-1)
SGML	Standard Generalized Markup Language

SGMP	Simple Gateway Management Protocol (IETF)
S-HTTP	Secure HTTP (IETF)
SIF	Signaling Information Field
SIG	Special Interest Group
SIO	Service Information Octet
SIR	Sustained Information Rate (SMDS)
SLA	Service level agreement
SLIP	Serial Line Interface Protocol (IETF)
SM	Switching Module
SMAP	System Management Application Part
SMDS	Switched Multimegabit Data Service
SMF	Single Mode Fiber
SMP	Simple Management Protocol
SMP	Switching Module Processor
SMPTE	Society of Motion Picture and Television Engineers
SMR	Specialized Mobile Radio
SMS	Standard Management System (SS7)
SMTP	Simple Mail Transfer Protocol (IETF)
SNA	Systems Network Architecture (IBM)
SNAP	Subnetwork Access Protocol
SNI	Subscriber Network Interface (SMDS)
SNMP	Simple Network Management Protocol (IETF)
SNP	Sequence Number Protection
SNR	Signtal-to-Noise Ratio
SONET	Synchronous Optical Network
SPAG	Standards Promotion and Application Group
SPARC	Scalable Performance Architecture
SPE	Synchronous payload envelope (SONET)
SPID	Service Profile Identifier (ISDN)
SPM	Self Phase Modulation
SPOC	Single Point of Contact
SPX	Sequenced Packet Exchange (NetWare)
SQL	Structured Query Language

SRB	Source Route Bridging
SRS	Stimulated Raman Scattering
SRT	Source Routing Transparent
SS7	Signaling System 7
SSL	Secure Socket Layer (IETF)
SSP	Service Switching Point (SS7)
SST	Spread Spectrum Transmission
STDM	Statistical Time Division Multiplexing
STM	Synchronous Transfer Mode
STM	Synchronous Transport Module (SDH)
STP	Signal Transfer Point (SS7)
STS	Synchronous Transport Signal (SONET)
STX	Start of Text (BISYNC)
SVC	Signaling virtual channel (ATM)
SVC	Switched Virtual Circuit
SXS	Step-by-Step Switching
SYN	Synchronization
SYNTRAN	Synchronous Transmission
TA	Terminal adapter (ISDN)
TAG	Technical Advisory Group
TASI	Time Assigned Speech Interpolation
TAXI	Transparent Asynchronous Transmitter/Receiver Interface (Physical Layer)
TCAP	Transaction Capabilities Application Part (SS7)
TCM	Time Compression Multiplexing
TCM	Trellis Coding Modulation
TCP	Transmission Control Protocol (IETF)
TDD	Time Division Duplexing
TDM	Time Division Multiplexing
TDMA	Time Division Multiple Access
TDR	Time Domain Reflectometer
TE1	Terminal Equipment type 1 (ISDN capable)
TE2	Terminal Equipment type 2 (non-ISDN capable)

TEI	Terminal Endpoint Identifier (LAPD)
TELRIC	Total Element Long-Run Incremental Cost
TIA	Telecommunications Industry Association
TIRKS	Trunk Integrated Record Keeping System
TL1	Transaction Language 1
TM	Terminal Multiplexer
TMN	Telecommunications Management Network
TMS	Time-Multiplexed Switch
TOH	Transport Overhead (SONET)
TOP	Technical and Office Protocol
TOS	Type of service (IP)
TP	Twisted Pair
TR	Token Ring
TRA	Traffic Routing Administration
TSI	Time Slot Interchange
TSLRIC	Total Service Long-Run Incremental Cost
TSO	Terminating Screening Office
TSO	Time-Sharing Option (IBM)
TSR	Terminate and Stay Resident
TSS	Telecommunication Standardization Sector (ITU-T)
TST	Time-Space-Time Switching
TSTS	Time-Space-Time-Space Switching
TTL	Time to Live
TU	Tributary Unit (SDH)
TUG	Tributary Unit Group (SDH)
TUP	Telephone User Part (SS7)
UA	Unnumbered Acknowledgment (HDLC)
UART	Universal Asynchronous Receiver Transmitter
UBR	Unspecified bit rate (ATM)
UDI	Unrestricted Digital Information (ISDN)
UDP	User Datagram Protocol (IETF)
UHF	Ultra High Frequency

UI	Unnumbered Information (HDLC)
UNI	User-to-Network Interface (ATM, FR)
UNIT™	Unified Network Interface Technology™ (Ocular)
UNMA	Unified Network Management Architecture
UPS	Uninterruptable Power Supply
UPSR	Unidirectional Path Switched Ring
UPT	Universal Personal Telecommunications
URL	Uniform Resource Locator
USART	Universal Synchronous Asynchronous Receiver Transmitter
UTC	Coordinated Universal Time
UTP	Unshielded Twisted Pair (Physical Layer)
UUCP	UNIX-UNIX Copy
VAN	Value-Added Network
VAX	Virtual Address Extension (DEC)
vBNS	Very High-speed Backbone Network Service
VBR	Variable Bit Rate (ATM)
VBR-NRT	Variable Bit Rate-Non-Real-Time (ATM)
VBR-RT	Variable Bit Rate-Real-Time (ATM)
VC	Virtual channel (ATM)
VC	Virtual circuit (PSN)
VC	Virtual container (SDH)
VCC	Virtual channel connection (ATM)
VCI	Virtual channel identifier (ATM)
VCSEL	Vertical Cavity Surface Emitting Laser
VDSL	Very High-speed Digital Subscriber Line
VDSL	Very High Bit-rate Digital Subscriber Line
VERONICA	Very Easy Rodent-Oriented Netwide Index to Computerized Archives (Internet)
VGA	Variable Graphics Array
VHF	Very High Frequency
VHS	Video Home System

VINES	Virtual Networking System (Banyan)
VIP	VINES Internet Protocol
VLF	Very Low Frequency
VLR	Visitor Location Register (Wireless/GSM)
VLSI	Very Large Scale Integration
VM	Virtual Machine (IBM)
VM	Virtual Memory
VMS	Virtual Memory System (DEC)
VOD	Video-on-Demand
VP	Virtual path
VPC	Virtual path connection
VPI	Virtual path identifier
VPN	Virtual private network
VR	Virtual reality
VSAT	Very Small Aperture Terminal
VSB	Vestigial Sideband
VSELP	Vector-Sum Excited Linear Prediction
VT	Virtual tributary
VTAM	Virtual Telecommunications Access Method (SNA)
VTOA	Voice and telephony over ATM
VTP	Virtual Terminal Protocol (ISO)
WACK	Wait Acknowledgment (BISYNC)
WACS	Wireless Access Communications System
WAIS	Wide Area Information Server (IETF)
WAN	Wide Area Network
WARC	World Administrative Radio Conference
WATS	Wide Area Telecommunications Service
WDM	Wavelength Division Multiplexing
WIN	Wireless In-building Network
WTO	World Trade Organization
WWW	World Wide Web (IETF)
WYSIWYG	What You See Is What You Get

xDSL	x-Type Digital Subscriber Line
XID	Exchange Identification (HDLC)
XNS	Xerox Network Systems
XPM	Cross Phase Modulation
YIQ	Y = luminance, I&Q are color
YUV	Y = luminance, U&V are color
ZBTSI	Zero Byte Time Slot Interchange
ZCS	Zero Code Suppression

APPENDIX C

Glossary

Abend   A contraction of the words *abnormal end* used to describe a computer crash in the mainframe world.

Absorption   A form of optical attenuation in which optical energy is converted into an alternative form, often heat. Often caused by impurities in the fiber, hydroxyl absorption is the best-known form.

Acceptance angle   The critical angle within which incident light is totally internally reflected inside the core of an optical fiber.

Access   The set of technologies used to reach the network by a user.

Active video   That part of the video signal that is visible.

Add-Drop Multiplexer (ADM)   A device used in SONET and SDH systems that has the capability to add and remove signal components without having to demultiplex the entire transmitted transmission stream, a significant advantage over legacy multiplexing systems such as DS-3.

Advanced Mobile Phone Service (AMPS)   The modern analog cellular network.

Aerial plant   Transmission equipment (including media, amplifiers, splice cases, and so on) that is suspended in the air between poles.

Alternate mark inversion   The encoding scheme used in T-1. Every other one is inverted in polarity from the one that preceded or follows it.

American Standards Code for Information Interchange (ASCII) A 7-bit data encoding scheme.

Amplifier   A device that increases the transmitted power of a signal. Amplifiers are typically spaced at carefully selected intervals along a transmission span.

Amplitude modulation   The process of causing an electromagnetic wave to carry information by changing or modulating the amplitude or loudness of the wave.

Analog   A signal that is continuously varying in time. Functionally, the opposite of digital.

Angular misalignment   The reason for loss that occurs at the fiber ingress point. If the light source is improperly aligned with the fiber's core, some of the incident light will be lost, leading to reduced signal strength.

Application-Specific Integrated Circuit (ASIC)   A specially designed IC created for a specific application.

Arithmetic Logic Unit (ALU)   The brain of a CPU chip.

Armor   The rigid, protective coating on some fiber cables that protects them from crushing and from chewing by rodents.

Aspect ratio   The ration of the width of the screen to its height.

Asynchronous   Data that is transmitted between two devices that do not share a common clock source.

ATM   Asynchronous Transfer Mode; one of the family of so-called fast packet technologies characterized by low error rates, high speed, and low cost. ATM is designed to connect seamlessly with SONET and SDH.

ATM Adaptation Layer (AAL)   In ATM, the layer responsible for matching the payload being transported to a requested QoS level by assigning an ALL Type that the network responds to.

Attenuation   The reduction in signal strength in optical fiber that results from absorption and scattering effects.

Avalanche Photodiode (APD)   An optical semiconductor receiver that has the capability to amplify weak, received optical signals by multiplying the number of received photons to intensify the strength of the received signal. APDs are used in transmission systems where receiver sensitivity is a critical issue.

Axis   The center line of an optical fiber.

Back scattering   The problem that occurs when light is scattered backward into the transmitter of an optical system. This impairment is analogous to an echo that occurs in copper-based systems.

Backward Explicit Congestion Notification (BECN)   A bit used in frame relay for notifying a device that it is transmitting too much information into the network and is therefore in violation of its service agreement with the switch.

Bandwidth   A measure of the number of bits per second that can be transmitted down a channel or the range of frequencies within which a transmission system operates.

Baseband   In signaling, any technique that uses digital signal representation.

Baud   The *signaling rate* of a transmission system. This is one of the most misunderstood terms in all of telecommunications. Often used synonymously with bits-per-second, baud usually has a very different meaning. By using multibit encoding techniques, a single signal can simultaneously represent multiple bits. Thus, the bit rate can be many times the signaling rate.

Beam splitter   An optical device used to direct a single signal in multiple directions through the use of a partially reflective mirror or some form of an optical filter.

Bell System Reference Frequency (BSRF)   In the early days of the Bell System, a single timing source in the Midwest provided a timing signal for all central office equipment in the country. This signal, delivered from a very expensive cesium clock source, was known as the BSRF. Today, GPS is used as the main reference clock source.

Bending loss   Loss that occurs when a fiber is bent far enough that its maximum allowable bend radius is exceeded. In this case, some of the light escapes from the waveguide, resulting in signal degradation.

Bend radius   The maximum degree to which a fiber can be bent before serious signal loss or fiber breakage occurs. Bend radius is one of the functional characteristics of most fiber products.

Bidirectional   A system that is capable of transmitting simultaneously in both directions.

Binary   A counting scheme that uses Base 2.

Bit rate   Bits-per-second.

Bluetooth   An open wireless standard designed to operate at a gross transmission level of 1 Mbps. Bluetooth is being positioned as a connectivity standard for personal area networks.

Bragg Grating   A device that relies on the formation of interference patterns to filter specific wavelengths of light from a transmitted signal. In optical systems, Bragg Gratings are usually created by wrapping a grating of the correct size around a piece of fiber that has been made photosensitive. The fiber is then exposed to strong ultraviolet light that passes through the grating, forming areas of high and low refractive indices. Bragg Gratings (or filters, as they are often called) are used for selecting certain wavelengths of a transmitted signal and are often used in optical switches, DWDM systems, and tunable lasers.

Brightness   The intensity of the video signal.

Broadband   Historically, broadband meant any signal that is faster than the ISDN Primary Rate (T1 or E1). Today, it means "big pipe"—in other words, a very high transmission speed. In signaling, the term means analog.

Buffer   A coating that surrounds optical fiber in a cable and offers protection from water, abrasion, and so on.

Building Integrated Timing Supply (BITS)   The central office device that receives the clock signal from GPS or another source and feeds it to the devices in the office it controls.

Bus   The parallel cable that interconnects the components of a computer.

Butt splice   A technique in which two fibers are joined end to end by fusing them with heat or optical cement.

Cable   An assembly made up of multiple optical or electrical conductors as well as other inclusions such as strength members, waterproofing materials, armor, and so on.

Cable assembly   A complete optical cable that includes the fiber itself and terminators on each end to make it capable of attaching to a transmission or receive device.

Cable plant   The entire collection of transmission equipment in a system, including the signal emitters, the transport media, the switching and multiplexing equipment, and the receive devices.

Cable vault   The subterranean room in a central office where cables enter and leave the building.

Call center   A room in which operators receive calls from customers.

Carrier Sense Multiple Access with Collision Detection (CSMA/CD)   The medium access scheme used in Ethernet LANs and characterized by an "if it feels good, do it" approach.

Cell   The standard protocol data unit in ATM networks. It comprises a 5-byte header and a 48-octet payload field.

Cell Loss Priority (CLP)   In ATM, a rudimentary single-bit field used to assign priority to transported payloads.

Cell Relay Service (CRS)   In ATM, the most primitive service offered by service providers, consisting of nothing more than raw bit transport with no assigned AAL types.

Cellular telephony   The wireless telephony system characterized by the following: low-power cells, frequency reuse, handoff, and central administration.

Center wavelength   The central operating wavelength of a laser used for data transmission.

Central office   A building that houses shared telephony equipment such as switches, multiplexers, and cable distribution hardware.

Central office terminal (COT)   In loop carrier systems, the device located in the central office that provides multiplexing and demultiplexing services. It is connected to the remote terminal.

Central Processing Unit (CPU)   Literally, the chipset in a computer that provides the intelligence.

Chained layers   The lower three layers of the OSI model that provide connectivity.

Chirp   A problem that occurs in laser diodes when the center wavelength shifts momentarily during the transmission of a single pulse. Chirp is due to instability of the laser itself.

Chromatic dispersion   Because the wavelength of transmitted light determines its propagation speed in an optical fiber, different wavelengths of light will travel at different speeds during transmission. As a result, the multiwavelength pulse will tend to spread out during transmission, causing difficulties for the receive device. Material dispersion, waveguide dispersion, and profile dispersion all contribute to the problem.

Chrominance   The color component of the NTSC or PAL signal.

Circuit Emulation Service (CES)   In ATM, a service that emulates private line service by modifying (1) the number of cells transmitted per second and (2) the number of bytes of data contained in the payload of each cell.

Cladding   The fused silica coating that surrounds the core of an optical fiber. It typically has a different index of refraction than the core, causing light that escapes from the core into the cladding to be refracted back into the core.

Coating   The plastic substance that covers the cladding of an optical fiber. It is used to prevent damage to the fiber itself through abrasion.

Code Division Multiple Access (CDMA)   One of several digital cellular access schemes. CDMA relies on frequency hopping and noise modulation to encode conversations.

Coherent   A form of emitted light in which all the rays of the transmitted light align themselves the same transmission axis, resulting in a narrow, tightly focused beam. Lasers emit coherent light.

Color space   Mathematical representation of color. YUV, YIQ, and RGB are examples of color space renderings.

Committed Information Rate (CIR)   The volume of data that a frame relay provider absolutely guarantees it will transport for a customer.

Competitive Local Exchange Carrier (CLEC)   A small telephone company that competes with the incumbent player in its own marketplace.

Complimentary Metal Oxide Semiconductor (CMOS)   A form of integrated circuit technology that is typically used in low-speed and low-power applications.

Component video   A video signal that keeps all components separate during transmission.

Composite video   A video signal that combines all signal components into a single transmission stream.

Compression   The process of reducing the size of a transmitted file without losing the integrity of the content by eliminating redundant information prior to transmitting or storing.

Concatenation   The technique used in SONET and SDH in which multiple payloads are grouped together to form a superrate frame capable of transporting payloads greater in size than the basic transmission speed of the system. Thus, an OC-12c provides 622.08 Mbps of total bandwidth, as opposed to an OC-12, which also offers 622.08 Mbps but in increments of OC-1 (51.84 Mbps).

Conditioning   The process of doctoring a dedicated circuit to eliminate the known and predictable results of distortion.

Congestion   The condition that results when traffic arrives faster than it can be processed by a server.

Connectivity   The process of providing electrical transport of data.

Connector   A device, usually mechanical, used to connect a fiber to a transmit or receive device or to bond two fibers.

Consultative Committee on International Telegraphy and Telephony (CCITT)   Now defunct and replaced by the ITU-TSS.

Core   The central portion of an optical fiber that provides the primary transmission path for an optical signal. It usually has a higher index of refraction than the cladding.

Counter-rotating ring   A form of transmission system that comprises two rings operating in opposite directions. Typically, one ring serves as the active path while the other serves as the protect or backup path.

Critical angle   The angle at which total internal reflection occurs.

Cross-Phase Modulation (XPM)   A problem that occurs in optical fiber that results from the nonlinear index of refraction of the silica in the fiber. Because the index of refraction varies according to the strength of the transmitted signal, some signals interact with each other in destructive ways. XPM is considered to be a fiber nonlinearity.

Cutoff wavelength   The wavelength below which single mode fiber ceases to be single mode.

Cyclic Redundancy Check (CRC)   A mathematical technique for checking the integrity of the bits in a transmitted file.

Cylinder   A stack of tracks to which data can be logically written on a hard drive.

Dark fiber   Optical fiber that is sometimes leased to a client that is not connected to a transmitter or receiver. In a dark fiber installation, it is the customer's responsibility to terminate the fiber.

Data   Raw, unprocessed zeroes and ones.

Data Circuit Terminating Equipment (DCE)   A modem or other device that delineates the end of the service provider's circuit.

Data communications   The science of moving data between two or more communicating devices.

Datagram   The service provided by a connectionless network. Often said to be unreliable, this service makes no guarantees with regard to latency or sequentiality.

Data Terminal Equipment (DTE)   User equipment that is connected to a DCE.

Decibel (dB)   Logarithmic measure of the strength of a transmitted signal. Because it is a logarithmic measure, a 20 dB loss would indicate that the received signal is one one-hundredth its original strength.

Dense Wavelength Division Multiplexing (DWDM)   A form of frequency division multiplexing in which multiple wavelengths of light are transmitted across the same optical fiber. These DWDM systems typically operate in the so-called L-Band (1625 nm) and have channels that are spaced between 50 and 100 GHz apart. Newly announced products may dramatically reduce this spacing.

Detector An optical receive device that converts an optical signal into an electrical signal so that it can be handed off to a switch, router, multiplexer, or other electrical transmission device. These devices are usually either NPN or APDs.

Diameter mismatch loss Loss that occurs when the diameter of a light emitter and the diameter of the ingress fiber's core are dramatically different.

Dichroic filter A filter that transmits light in a wavelength-specific fashion, reflecting nonselected wavelengths.

Dielectric A substance that is nonconducting.

Diffraction Grating A grid of closely spaced lines that are used to selectively direct specific wavelengths of light as required.

Digital A signal characterized by discrete states. The opposite of analog.

Digital hierarchy In North America, the multiplexing hierarchy that enables 64 Kbps DS-0 signals to be combined to form DS-3 signals for high bit rate transport.

Digital signal level 0 (DS-0) A 64 Kbps signal.

Digital signal level 1 (DS-1) A 1.544 Mbps signal.

Digital signal level 2 (DS-2) A 6.312 Mbps signal.

Digital Subscriber Line (DSL) A technique for transporting high-speed digital data across the analog local loop while (in some cases) transporting voice simultaneously.

Digital Subscriber Line Access Multiplexer (DSLAM) The multiplexer in the central office that receives voice and data signals on separate channels, relaying voice to the local switch and data to a router elsewhere in the office.

Diode A semiconductor device that only enables the current to flow in a single direction.

Discard Eligibility bit (DE) A primitive single-bit technique for prioritizing traffic that is to be transmitted.

Dispersion The spreading of a light signal over time that results from modal or chromatic inefficiencies in the fiber.

Dispersion Compensating Fiber (DCF) A segment of fiber that exhibits the opposite dispersion effect of the fiber to which it is coupled. DCF is used to counteract the dispersion of the other fiber.

Dispersion-Shifted Fiber (DSF) A form of optical fiber that is designed to exhibit zero dispersion within the C-Band (1550 nm). DSF does not work well for DWDM because of Four Wave Mixing problems; Non-Zero Dispersion-Shifted Fiber is used instead.

Distortion   A known and measurable (and therefore correctable) impairment on transmission facilities.

Dopant   Substances used to lower the refractive index of the silica used in optical fiber.

DS3   A 44.736 Mbps signal format found in the North American Digital Hierarchy.

Dual-Tone, Multifrequency (DTMF)   The set of tones used in modern phones to signal dialed digits to the switch. Each button triggers a pair of tones.

Duopoly   The current regulatory model for cellular systems; two providers are assigned to each market. One is the wireline provider (typically the local ILEC), the other an independent provider.

E1   The 2.048 Mbps transmission standard found in Europe and other parts of the world. It is analogous to the North American T1.

Edge-emitting diode   A diode that emits light from the edge of the device rather than the surface, resulting in a more coherent and directed beam of light.

Effective area   The cross-section of a single-mode fiber that carries the optical signal.

Encryption   The process of modifying a text or image file to prevent unauthorized users from viewing the content.

End-to-end layers   The upper four layers of the OSI model that provide interoperability.

Enhanced Data for Global Evolution (EDGE)   A 384 Kbps enhancement to GSM.

Erbium-Doped Fiber Amplifier (EDFA)   A form of optical amplifier that uses the element erbium to bring about the amplification process. Erbium has the enviable quality that, when struck by light operating at 980 nm, it emits photons in the 1550 nm range, thus providing agnostic amplification for signals operating in the same transmission window.

Ethernet   A LAN product developed by Xerox that relies on a CSMA/CD medium access scheme.

Evanescent wave   Light that travels down the inner layer of the cladding instead of down the fiber core.

Excess Information Rate (EIR)   The amount of data that is being transmitted by a user above the CIR in frame relay.

Extended Binary Coded Decimal Interchange Code (EBCDIC)   An 8-bit data encoding scheme.

Extended Superframe (ESF)   The framing technique used in modern T-carrier systems that provides a dedicated data channel for nonintrusive testing of customer facilities.

Extrinsic loss   Loss that occurs at splice points in an optical fiber.

Eye pattern   A measure of the degree to which bit errors are occurring in optical transmission systems. The width of the "eyes" (eye patterns look like figure eights lying on their sides) indicates the relative bit error rate.

Facility   A circuit.

Faraday Effect   Sometimes called the magneto-optical effect, the Faraday Effect describes the degree to which some materials can cause the polarization angle of incident light to change when placed within a magnetic field that is parallel to the propagation direction.

Fast packet   Technologies characterized by low error rates, high speed, and low cost.

Ferrule   A rigid or semirigid tube that surrounds optical fibers and protects them.

Fiber grating   A segment of photosensitive optical fiber that has been treated with ultraviolet light to create a refractive index within the fiber that varies periodically along its length. It operates analogously to a fiber grating and is used to select specific wavelengths of light for transmission.

Fiber-to-the-Curb (FTTC)   A transmission architecture for service delivery in which a fiber is installed in a neighborhood and terminated at a junction box. From there, coaxial cable or twisted pair can be cross-connected from the O-E converter to the customer premises. If coax is used, the system is called *Hybrid Fiber Coax* (HFC); twisted pair-based systems are called *Switched Digital Video* (SDV).

Fiber-to-the-Home (FTTH)   Similar to FTTC except that FTTH extends the optical fiber all the way to the customer premises.

Field   Either of two components of a television signal. One field contains the odd-numbered scan lines, while the other contains the even-numbered lines. Together they make up a frame.

Forward Error Correction (FEC)   An error correction technique that sends enough additional overhead information along with the transmitted data that a receiver can not only detect an error, but actually fix it without requesting a resend.

Forward Explicit Congestion Notification (FECN)   A bit in the header of a frame relay frame that can be used to notify a distant switch that the frame experienced severe congestion on its way to the destination.

Four Wave Mixing (FWM)   The nastiest of the so-called fiber nonlinearities. FWM is commonly seen in DWDM systems and occurs when the closely spaced channels mix and generate the equivalent of optical sidebands. The number of these sidebands can be expressed by the equation, sidebands = $1/2 \ (n^3 - n^2)$, where n is the number of original channels in the system. Thus, a 16-channel DWDM system will potentially generate 1,920 interfering sidebands!

Frame   A variable size data transport entity.

Frame rate   The rate at which frames are transmitted. In video the rate is 30 frames per second; in film it is 24 frames per second.

Frame relay   One of the family of so-called fast packet technologies characterized by low error rates, high speed, and low cost.

Frame Relay Bearer Service (FRBS)   In ATM, a service that enables a frame relay frame to be transported across an ATM network.

Frequency-agile   The capability of a receiving or transmitting device to change its frequency in order to take advantage of alternate channels.

Frequency Division Multiple Access (FDMA)   The access technique used in analog AMPS cellular systems.

Frequency division multiplexing (FDM)   The process of assigning specific frequencies to specific users.

Frequency modulation   The process of causing an electromagnetic wave to carry information by changing or modulating the frequency of the wave.

Fresnel loss   The loss that occurs at the interface between the head of the fiber and the light source to which it is attached. At air-glass interfaces, the loss usually equates to about 4 percent.

Full-duplex   Two-way simultaneous transmission.

Fused fiber   A group of fibers that are fused together so that they will remain in alignment. They are often used in one-to-many distribution systems for the propagation of a single signal to multiple destinations. Fused fiber devices play a key role in *passive optical networking* (PON).

Fusion splice   A splice made by melting the ends of the fibers together.

General Packet Radio Service (GPRS)   Another add-on for GSM networks that is not enjoying a great deal of success in the market yet. Stay tuned.

Generic Flow Control (GFC)   In ATM, the first field in the cell header. It is largely unused except when it is overwritten in NNI cells, in which case it becomes additional space for virtual path addressing.

Geosynchronous Earth Orbit Satellite (GEOS)   A family of satellites that orbit above the equator at an altitude of 22,300 miles and provide data and voice transport services.

Global Positioning System (GPS)   The array of satellites used for radiolocation around the world. In the telephony world, GPS satellites provide an accurate timing signal for synchronizing office equipment.

Global System for Mobile Communications (GSM)   The wireless access standard used in many parts of the world that offers two-way paging, short messaging, and two-way radio in addition to cellular telephony.

Go-Back-N   A technique for error correction that causes all frames of data to be transmitted again, starting with the errored frame.

Gozinta   "Goes into."

Gozouta   "Goes out of."

Graded Index Fiber (GRIN)   A type of fiber in which the refractive index changes gradually between the central axis of the fiber and the outer layer, instead of abruptly at the core-cladding interface.

Graphical User Interface (GUI)   The computer interface characterized by the "click, move, drop" method of file management.

Groom and fill   Similar to add-drop, groom and fill refers to the capability to add (fill) and drop (groom) payload components at intermediate locations along a network path.

H.261   Video compression standard created for videoconferencing.

H.323   A videoconferencing standard created for the transmission of videoconferencing content over ISDN lines.

Half-duplex   Two-way transmission but only one direction at a time.

Haptics   The science of providing tactile feedback to a user electronically. Often used in high-end virtual reality systems.

Headend   The signal origination point in a cable system.

Header   In ATM, the first five bytes of the cell. The header contains information used by the network to route the cell to its ultimate destination. Fields in the cell header include Generic Flow Control, Virtual Path Identifier, Virtual Channel Identifier, Payload Type Identifier, Cell Loss Priority, and Header Error Correction.

Header Error Correction (HEC)   In ATM, the header field used to recover from bit errors in the header data.

Hertz (Hz)   The measure of cycles per second in transmission systems.

Hop count   A measure of the number of machines a message or packet has to pass through between the source and the destination. Often used as a diagnostic tool.

Hybrid fiber coax   A transmission system architecture in which a fiber feeder penetrates a service area and is then cross-connected to coaxial cable feeders into the customers' premises.

Incumbent Local Exchange Carrier (ILEC)   A Regional Bell Operating Company (RBOC).

Index of refraction   A measure of the ratio between the velocity of light in a vacuum and the velocity of the same light in an optical fiber. The refractive index is always greater than one and is denoted n.

Information   Data that has been converted to manipulatable form.

Infrared (IR)   The region of the spectrum within which most optical transmission systems operate, found between 700 nm and 0.1 mm.

Injection laser   A semiconductor laser (synonym).

Inside plant   Telephony equipment that is outside of the central office.

Integrated Services Digital Network (ISDN)   A digital local loop technology that offers moderately high bit rates to customers.

Intensity   Rightness.

Interlaced   A screen in which two fields are used to create a single frame of video.

Intermodulation   A fiber nonlinearity that is similar to four-wave mixing, in which the power-dependent refractive index of the transmission medium enables signals to mix and create destructive sidebands.

International Telecommunications Union (ITU)   A division of the United Nations that is responsible for managing the telecomm standards development and maintenance processes.

Internet Service Provider (ISP)   A company that offers Internet access.

Interoperability   In SONET and SDH, the capability of devices from different manufacturers to send and receive information to and from each other successfully.

Intrinsic loss   Loss that occurs as the result of physical differences in the two fibers being spliced.

Isochronous   A word used in timing systems that means that there is constant delay across a network.

ITU Telecommunications Standardization Sector (ITU-TSS)   The ITU organization responsible for telecommunications standards development.

Jacket   The protective outer coating of an optical fiber cable. The jacket may be polyethylene, Kevlar®, or metallic.

Joint Photographic Experts Group (JPEG)   A standards body tasked with developing standards for the compression of still images.

Jumper   An optical cable assembly, usually fairly short, that is terminated on both ends with connectors.

Knowledge   Information that has been acted upon and modified through some form of intuitive human thought process.

LAN Emulation (LANE)   In ATM, a service that defines the capability to provide bridging services between LANs across an ATM network.

Large core fiber   Fiber that characteristically has a core diameter of 200 microns or more.

Laser Diode (LD)   A diode that produces coherent light when a forward biasing current is applied to it.

Light Amplification by the Stimulated Emission of Radiation (Laser)   Lasers are used in optical transmission systems because they produce coherent light that is almost purely monochromatic.

Light Emitting Diode (LED)   A diode that emits incoherent light when a forward bias current is applied to it. LEDs are typically used in shorter-distance, lower-speed systems.

Lightguide   A term that is used synonymously with optical fiber.

Line Overhead (LOH)   In SONET, the overhead that is used to manage the network regions between multiplexers.

Linewidth   The spectrum of wavelengths that make up an optical signal.

Load coil   A device that tunes the local loop to the voiceband.

Local Access and Transport Area (LATA)   The geographic area within which an ILEC is enabled to transport traffic. Beyond LATA boundaries, the ILEC must hand traffic off to a long-distance carrier.

Local Area Network (LAN)   A small network that has the following characteristics: privately owned, high speed, low error rate, physically small.

Local loop   The pair of wires (or digital channel) that runs between the customer's phone (or computer) and the switch in the local central office.

Loose tube optical cable   An optical cable assembly in which the fibers within the cable are loosely contained within tubes inside the sheath of the cable. The fibers are able to move within the tube, thus enabling them to adapt and move without damage as the cable is flexed and stretched.

Loss   The reduction in signal strength that occurs over distance, usually expressed in decibels.

Low Earth Orbit Satellite (LEOS)   Satellites that orbit pole to pole instead of above the equator and offer near-instantaneous response time.

M13   A multiplexer that interfaces between DS-1 and DS-3 systems.

Main Distribution Frame (MDF)   The large iron structure that provides physical support for cable pairs in a central office between the switch and the incoming/outgoing cables.

Mainframe   A large computer that offers support for very large databases and large numbers of simultaneous sessions.

Material dispersion   A dispersion effect caused by the fact that different wavelengths of light travel at different speeds through a medium.

Message switching   An older technique that sends entire messages from point to point instead of breaking the message into packets.

Metasignaling Virtual Channel (MSVC)   In ATM, a signaling channel that is always on. It is used for the establishment of temporary signaling channels as well as channels for voice and data transport.

Metropolitan Area Network (MAN)   A network, larger than a LAN, that provides high-speed services within a metropolitan area.

Microbend   Changes in the physical structure of an optical fiber caused by bending that can result in light leakage from the fiber.

Midspan meet   In SONET and SDH, the term used to describe interoperability. *See* interoperability.

Mobile Telephone Switching Office (MTSO)   A central office with special responsibilities for handling cellular services and the interface between cellular users and the wireline network.

Modal dispersion   *See* Multimode dispersion.

Mode   A single wave that propagates down a fiber. Multimode fiber enables multiple modes to travel, while single mode fiber enables only a single mode to be transmitted.

Modem   A term from the words *modulate* and *demodulate*. Its job is to make a computer appear to the network like a telephone.

Modulation   The process of changing or *modulating* a carrier wave to cause it to carry information.

Moving Picture Experts Group (MPEG)   A standards body tasked with crafting standards for motion pictures.

Multimode dispersion   Sometimes referred to as modal dispersion, multimode dispersion is caused by the fact that different modes take different times to move from the ingress point to the egress point of a fiber, thus resulting in modal spreading.

Multimode fiber   Fiber that has a core diameter of 62.5 microns or greater, wide enough to enable multiple modes of light to be simultaneously transmitted down the fiber.

Multiplexer   A device that has the capability to combine multiple inputs into a single output as a way to reduce the requirement for additional transmission facilities.

Multiprotocol over ATM (MPOA)   In ATM, a service that enables IP packets to be routed across an ATM network.

National Television Systems Committee (NTSC)   The standard for television in the United States.

Near-End Crosstalk (NEXT)   The problem that occurs when an optical signal is reflected back toward the input port from one or more output ports. This problem is sometimes referred to as *isolation directivity*.

Noise   An unpredictable impairment in networks. It cannot be anticipated; it can only be corrected after the fact.

Non-Dispersion-Shifted Fiber (NDSF)   Fiber that is designed to operate at the low-dispersion second operational window (1310 nm).

Non-Zero Dispersion-Shifted Fiber (NZDSF)   A form of single mode fiber that is designed to operate just outside the 1550 nm window so that fiber nonlinearities, particularly FWM, are minimized.

Numerical Aperture (NA)   A measure of the capability of a fiber to gather light, NA is also a measure of the maximum angle at which a light source can be from the center axis of a fiber in order to collect light.

Operations, Administration, Maintenance and Provisioning (OAM&P)   The four key areas in modern network management systems. OAM&P was first coined by the Bell System and continues in widespread use today.

Operations Support Systems (OSS)   Another term for OAM&P.

Optical amplifier   A device that amplifies an optical signal without first converting it to an electrical signal.

Optical Carrier level n (OC-n)   A measure of bandwidth used in SONET systems. OC-1 is 51.84 Mbps; OC-n is n times 51.84 Mbps.

Optical isolator   A device used to selectively block specific wavelengths of light.

Optical Time Domain Reflectometer (OTDR)   A device used to detect failures in an optical span by measuring the amount of light reflected back from the air-glass interface at the failure point.

Outside plant   Telephone equipment that is outside of the central office.

Overhead   The part of a transmission stream that the network uses to manage and direct the payload to its destination.

Packet   A variable size entity normally carried inside a frame or cell.

Packet switching   The technique for transmitting packets across a wide area network.

Path overhead   In SONET and SDH, the part of the overhead that is specific to the payload being transported.

Payload   In SONET and SDH, the user data that is being transported.

Payload Type Identifier (PTI)   In ATM, a cell header field that is used to identify network congestion and cell type. The first bit indicates whether the cell was generated by the user or by the network, while the second indicates the presence or absence of congestion in user-generated cells or flow-related *Operations, Administration, & Maintenance* (OA&M) information in cells generated by the network. The third bit is used for service-specific, higher-layer functions in the user-to-network direction, such as to indicate that a cell is the last in a *series* of cells. From the network to the user, the third bit is used with the second bit to indicate whether the OA&M information refers to segment or end-to-end-related information flow.

Permanent Virtual Circuit (PVC)   A circuit provisioned in frame relay or ATM that does not change without service order activity by the service provider.

Phased Alternate Line (PAL)   The television standard in Europe.

Phase modulation   The process of causing an electromagnetic wave to carry information by changing or modulating the phase of the wave.

Photodetector   A device used to detect an incoming optical signal and convert it to an electrical output.

Photodiode   A semiconductor that converts light to electricity.

Photon   The fundamental unit of light, sometimes referred to as a quantum of electromagnetic energy.

Photonic   The optical equivalent of the term *electronic*.

Pipelining   The process of having multiple unacknowledged outstanding messages in a circuit between two communicating devices.

Pixel   Contraction of the term *picture element*. The tiny color elements that make up the screen on a computer monitor.

Planar waveguide   A waveguide fabricated from a flat material, such as a sheet of glass, into which are etched fine lines used to conduct optical signals.

Plenum   The air-handling space in buildings found inside walls, under floors, and above ceilings. The plenum spaces are often used as conduits for optical cables.

Plenum cable   Cable that passes fire retardant tests so that it can legally be used in plenum installations.

Plesiochronous   In timing systems, a term that means "almost synchronized." It refers to the fact that in SONET and SDH systems, payload components frequently derive from different sources and therefore may have slightly different phase characteristics.

Pointer   In SONET and SDH, a field that is used to indicate the beginning of the transported payload.

Polarization   The process of modifying the direction of the magnetic field within a light wave.

Polarization Mode Dispersion (PMD)   The problem that occurs when light waves with different polarization planes in the same fiber travel at different velocities down the fiber.

Preform   The cylindrical mass of highly pure fused silica from which optical fiber is drawn during the manufacturing process. In the industry, the preform is sometimes referred to as a gob.

Private Branch Exchange (PBX)   Literally, a small telephone switch located on a customer premises. The PBX connects back to the service provider's central office via a collection of high-speed trunks.

Private line   A dedicated point-to-point circuit.

Protocol   A set of rules that facilitates communications.

Pulse Code Modulation (PCM)   The encoding scheme used in North America for digitizing voice.

Pulse spreading   The widening or spreading out of an optical signal that occurs over distance in a fiber.

Pump laser   The laser that provides the energy used to excite the dopant in an optical amplifier.

Q.931   The set of standards that defines signaling packets in ISDN networks.

Quantize   The process of assigning numerical values to the digitized samples created as part of the voice digitization process.

Random Access Memory (RAM)   The volatile memory used in computers for short-term storage.

Rayleigh Scattering   A scattering effect that occurs in optical fiber as the result of fluctuations in silica density or chemical composition. Metal ions in the fiber often cause Rayleigh Scattering.

Read-Only Memory (ROM)   Memory that cannot be erased; often used to store critical files or boot instructions.

Refraction   The change in direction that occurs in a light wave as it passes from one medium into another. The most common example is the bending that is often seen to occur when a stick is inserted into water.

Refractive index   A measure of the speed at which light travels through a medium, usually expressed as a ratio compared to the speed of the same light in a vacuum.

Regenerative repeater   A device that reconstructs and regenerates a transmitted signal that has been weakened over distance.

Regenerator   A device that recreates a degraded digital signal before transmitting it onto its final destination.

Regional Bell Operating Company (RBOC)   Today called an Incumbent Local Exchange Carrier (ILEC).

Remote Terminal (RT)   In loop carrier systems, the multiplexer located in the field. It communicates with the *central office terminal* (COT).

Repeater   *See* regenerator.

Scattering   The backsplash or reflection of an optical signal that occurs when it is reflected by small inclusions or particles in the fiber.

Section Overhead (SOH)   In SONET systems, the overhead that is used to manage the network regions that occur between repeaters.

Sector   A quadrant on a disk drive to which data can be written. Used for locating information on the drive.

Selective retransmit   An error correction technique in which only the errored frames are retransmitted.

Self-Phase Modulation (SPM)   The refractive index of glass is directly related to the power of the transmitted signal. As the power fluctuates, so too does the index of refraction, causing waveform distortion.

Sequential Color with Memory (SECAM)   A television broadcast standard used in many parts of the world.

Sheath   One of the layers of protective coating in an optical fiber cable.

Signaling   The techniques used to set up, maintain, and tear down a call.

Signaling System 7 (SS7)   The current standard for telephony signaling worldwide.

Signaling Virtual Channel (SVC)   In ATM, a temporary signaling channel used to establish paths for the transport of user traffic.

Simplex   One way transmission *only*.

Single Mode Fiber (SMF)   The most popular form of fiber today, characterized by the fact that it enables only a single mode of light to propagate down the fiber.

Soliton   A unique waveform that takes advantage of nonlinearities in the fiber medium, the result of which is a signal that suffers essentially no dispersion effects over long distances. Soliton transmission is an area of significant study at the moment because of the promise it holds for long-haul transmission systems.

Source   The emitter of light in an optical transmission system.

Standards   The published rules that govern an industry's activities.

Step index fiber   Fiber that exhibits a continuous refractive index in the core, which then steps at the core-cladding interface.

Stimulated Brillouin Scattering (SBS)   A fiber nonlinearity that occurs when a light signal traveling down a fiber interacts with acoustic vibrations in the glass matrix (sometimes called *photon-phonon interaction*), causing light to be scattered or reflected back toward the source.

Stimulated Raman Scattering (SRS)   A fiber nonlinearity that occurs when power from short-wavelength, high-power channels is bled into longer-wavelength, lower-power channels.

Store-and-forward   The transmission technique in which data is transmitted to a switch, stored there, examined for errors, examined for address information, and forwarded onto the final destination.

Strength member   The strand within an optical cable that is used to provide tensile strength to the overall assembly. The member is usually composed of steel, fiberglass, or Aramid yarn.

Surface emitting diode   A semiconductor that emits light from its surface, resulting in a low-power, broad-spectrum emission.

SVC   A frame relay or ATM technique in which a customer can establish on-demand circuits as required.

Synchronous   A term that means that both communicating devices derive their synchronization signal from the same source.

Synchronous Digital Hierarchy (SDH)   The European equivalent of SONET.

Synchronous Optical Network (SONET) A multiplexing standard that begins at DS-3 and provides standards-based multiplexing up to gigabit speeds. SONET is widely used in telephone company long-haul transmission systems and was one of the first widely deployed optical transmission systems.

Synchronous Transmission Signal Level 1 (STS-1) In SONET systems, the lowest transmission level in the hierarchy. STS is the electrical equivalent of OC.

T1 The 1.544 Mbps transmission standard in North America.

T-3 In the North American Digital Hierarchy, a 44.736 Mbps signal.

Tandem A switch that serves as an interface between other switches and typically does not directly host customers.

Telecommunications The science of transmitting sound over distance.

Terminal multiplexer In SONET and SDH systems, a device that is used to distribute payload to or receive payload from user devices at the end of an optical span.

Tight buffer cable An optical cable in which the fibers are tightly bound by the surrounding material.

Time Division Multiple Access (TDMA) A digital technique for cellular access in which customers share access to a frequency on a round-robin, time-division basis.

Time-division multiplexing (TDM) The process of assigning time slots to specific users.

Total internal reflection The phenomenon that occurs when light strikes a surface at such an angle that all of the light is reflected back into the transporting medium. In optical fiber, total internal reflection is achieved by keeping the light source and the fiber core oriented along the same axis so that the light that enters the core is reflected back into the core at the core-cladding interface.

Transceiver A device that incorporates both a transmitter and a receiver in the same housing, thus reducing the need for rack space.

Transponder A device that incorporates a transmitter, a receiver, and a multiplexer on a single chassis.

Twisted pair The wire used to interconnect customers to the telephone network.

Uninterruptible Power Supply (UPS)   Part of the central office power plant that prevents power outages.

Vertical Cavity Surface Emitting Laser (VCSEL)   A small, highly efficient laser that emits light vertically from the surface of the wafer on which it is made.

Virtual Channel (VC)   In ATM, a unidirectional channel between two communicating devices.

Virtual Channel Identifier (VCI)   In ATM, the field that identifies a virtual channel.

Virtual container   In SDH, the technique used to transport subrate payloads.

Virtual Path (VP)   In ATM, a combination of unidirectional virtual channels that make up a bidirectional channel.

Virtual Path Identifier (VPI)   In ATM, the field that identifies a virtual path.

Virtual private network   A network connection that provides private-like services over a public network.

Virtual Tributary (VT)   In SONET, the technique used to transport subrate payloads.

Voice/Telephony over ATM (VTOA)   In ATM, a service used to transport telephony signals across an ATM network.

Voiceband   The 300 to 3300 Hz band used for the transmission of voice traffic.

Waveguide   A medium that is designed to conduct light within itself over a significant distance, such as optical fiber.

Waveguide dispersion   A form of chromatic dispersion that occurs when some of the light traveling in the core escapes into the cladding, traveling there at a different speed than the light in the core.

Wavelength   The distance between the same points on two consecutive waves in a chain—for example, from the peak of wave one to the peak of wave two. Wavelength is related to frequency by the equation where $(\lambda = \frac{c}{f})$ lambda (λ) is the wavelength, c is the speed of light, and f is the frequency of the transmitted signal.

Wavelength Division Multiplexing (WDM)   The process of transmitting multiple wavelengths of light down a fiber.

Wide Area Network (WAN)   A network that provides connectivity over a large geographical area.

Window   A region within which optical signals are transmitted at specific wavelengths to take advantage of propagation characteristics that occur there, such as minimum loss or dispersion.

Window size   A measure of the number of messages that can be outstanding at any time between two communicating entities.

YIQ   The color space used in NTSC television systems.

YUV   The color space used in PAL television systems. It can also be used in NTSC systems.

Zero dispersion wavelength   The wavelength at which material and waveguide dispersion cancel each other.

APPENDIX D

Videoconferencing Bibliography

2600 Magazine: The Hacker Quarterly. Various issues.

Aukstakalnis, Steve and David Blatner. "Silicon Mirage: The Art and Science of Virtual Reality," Peach Pit Press, 1992, ISBN 0-938151-82-7.

Bell Laboratories Editors. *Engineering and Operations in the Bell System.* R. F. Rey, Technical Editor, prepared by members of the technical staff and the Technical Publication Department, AT&T Bell Laboratories. Murray Hill, New Jersey, 1983.

Benedikt, Michael. "Cyberspace—First Steps," MIT Press, 1992 (collection of essays on VR), ISBN 0-262-52177-6.

Berners-Lee, Tim. *Information Management: A Proposal.* CERN, May, 1990.

Brooks, John. *Telephone.* New York: Harper & Row, 1976.

Burke, James. *Connections.* New York: Little-Brown, 1978.

Clarke, Arthur C. *How the World Was One: Beyond the Global Village.* New York: Bantam, 1992.

Cohen, Frederick. *Protection and Security on the Information Superhighway.* New York: John Wiley and Sons, 1995.

Davis, Stanley M., Christopher Meyer, and Stan Davis. *Blur: The Speed of Change in the Connected Economy.* Reading, MA: Addison-Wesley, 1998.

Diamond, Jared. *Guns, Germs, and Steel.* New York: W. W. Norton, 1999.

Dibbell, Julian. "The Prisoner: Phiber Optik Goes Directly to Jail." *Village Voice,* January 11, 1994.

Earnshaw, R. A., M.A. Gigante, and H. Jones. *Virtual Reality Systems.* April 1993, ISBN: 0 12 227748.

Ellis, Stephen R. *Pictorial Communication in Virtual and Real Environments.* Taylor and Francis. 1991, ISBN: 0-74840008-7.

Evans, Philip and Thomas S. Wurster. *Blown to Bits: How the New Economics of Information Transforms Strategy.* Boston, MA: Harvard Business School Press, 2000.

Flanagan, William A. *The Guide to T-1 Networking: How to Buy, Install and Use T-1, from Desktop to DS-3*. New York: Telecom Library, Inc, 1990.

Freedman, Thomas L. *The Lexus and the Olive Tree*. New York: Anchor Books, 2000.

Gibson, William and Bruce Sterling. *Speeches at National Academy of Sciences Convocation on Technology and Education*. Washington D.C., May 10, 1993.

Goff, David R. *Fiber Optic Reference Guide: A Practical Guide to the Technology—Second Edition*. Boston: Focal Press, 1999.

Goralski, Walter J. *ADSL and DSL Technologies*. New York: McGraw-Hill, 1998.

———. *SONET: A Guide to Synchronous Optical Networks*. New York: McGraw-Hill, 1997.

Goralski, Walter J. and Matthew C. Kolon. *IP Telephony*. New York: McGraw-Hill, 2000.

Gralla, Preston. *How the Internet Works*. Indianapolis: Macmillan, 1999.

Gregory, R. L. *Eye and Brain. The Psychology of Seeing (Fourth Edition)*. Weidenfeld and Nicholson. 1990, ISBN: 29782042-7.

Hafner, Katie and John Markoff. *Cyberpunk: Outlaws and Hackers on the Computer Frontier*. New York: Simon and Schuster, 1991.

Hamit, Francis. *Virtual Reality and the Exploration of Cyberspace*. Sams Publishing, ISBN 0-672-30361-2 (includes PC disk and a very extensive bibliography).

Hayward, Tom. *Adventures in Virtual Reality*. Que Books, 1993, ISBN 1-56529-208-1 (includes PC disk with VREAM world and other demos).

Hecht, Jeff. *Understanding Fiber Optics—Third Edition*. Upper Saddle River, NJ: Prentice-Hall, 1999.

Heim, Michael. *The Metaphysics of Virtual Reality*. Oxford University Press, 1993. ISBN 0-19-508178-1.

Heisle, Sandra and Judith Roth. *Virtual Reality: Theory, Practice, and Promise*. Meckler Corp., 1990, LC Call number BD331.V57 1991.

Holzmann, Gerard J. and Bjorn Pehrson. *The Early History of Data Networks*. Los Alamitos, CA: IEEE Computer Society Press, 1995.

Jacobson, Linda (Editor). *CYBERARTS: Exploring Art & Technology*. Miller Freeman, Inc., ISBN 0-87930-253-4.

————. *Garage Virtual Reality*. Sams Publishing, 1993. ISBN 0-672-30389-2 (A how-to book for the home brew enthusiast and includes PC disk).

Kartalopoulos, Stamatios. *Understanding SONET/SDH and ATM*. New York: IEEE Press, 1999.

Kessler, Gary. *An Overview of Cryptography*. Available from http://www.garykessler.net/library/crypto.html.

————. *Computer and Network Security.* Hill Associates Monograph Series, June 1995.

Krueger, Myron. *Artificial Reality II*. Reading, MA: Addison-Wesley, 1991. ISBN: 0-201-52260-8.

Lampton, Chris. *Flights of Fantasy, Programming 3-D Video Games in C++*. Waite Group, 1993, ISBN 1-878739-18.

Lanning, Michael J. *Delivering Profitable Value*. New York: Perseus Books, 1998.

Lavroff, Nicholas. *Virtual Reality Playhouse*. Waite Group Press, 1992, ISBN 1-878739-19-0 (includes PC disk, apps are with most WoW interactive animations).

Lubar, Steven. *Info Culture*. New York: Houghton Mifflin, 1993.

McCartney, Scott. *ENIAC*. New York: Walker, 1999.

Metz, Christopher Y. *IP Switching Protocols and Architectures*. New York: McGraw-Hill, 1999.

Minoli, Daniel. *Telecommunications Technology Handbook*. Norwood, MA: Artech House, 1991.

Minoli, Daniel and Emma Minoli. *Delivering Voice Over IP Networks*. New York: John Wiley & Sons, 1998.

Murray, Janet. *Hamlet on the Holodeck*. New York: Simon & Schuster, 1997.

Negroponte, Nicholas. *Being Digital*. New York: Alfred A. Knopf, 1995.

Newton, Harry. *Newton's Telecom Dictionary, 15th Edition*. New York: Miller-Freeman,1999.

Peters, Tom. *The Circle of Innovation: You Can't Shrink Your Way to Greatness*. New York: Vintage, 1999.

Pimentel, Ken and Kevin Teixeira. *Virtual Reality: Through the New Looking Glass*. New York: McGraw-Hill, 1993, ISBN 0-8306-4064-9.

Pomeranz, Kenneth and Steven Topik. *The World That Trade Created*. New York: M. E. Sharpe, 1999.

Porter, Michael. *Competitive Advantage: Creating and Sustaining Superior Performance*. Boston: Free Press, 1985.

————. *The Competitive Advantage of Nations*. New York: The Free Press, 1990.

Rackham, Neil. *Spin Selling*. New York: McGraw-Hill, 1998.

Rheingold, Howard. *Virtual Reality*. Summit Books, 1991, ISBN 0-671-69363-8.

Schein, Edgar H. *Organizational Psychology, Third Edition*. New Jersey: Prentice Hall, 1980.

Segaller, Stephen. *Nerds 2.0.1*. New York: TV Books, 1999.

Shannon, Claude. "A Mathematical Theory of Communication." Reprinted with corrections from the *Bell System Technical Journal,* Volume 27, pp. 379–423, 623–656, July and October 1948. Available at http://cm.bell-labs.com/cm/ms/what/shannonday/shannon1948.pdf.

Shepard, Steven. *Optical Networking Crash Course*. New York: McGraw-Hill, 2001.

————. *SONET and SDH Demystified*. New York: McGraw-Hill, 2001.

————. *Telecom Crash Course*. New York: McGraw-Hill, 2001.

————. *Telecommunications Convergence: How to Profit from the Convergence of Technologies, Services and Companies*. New York: McGraw-Hill, 2000.

Stallings, William. *Data and Computer Communications, Third Edition*. New York: Macmillan, 1991.

Stampe, Dave, Bernie Roehl, and John Eagan. *Virtual Reality Creations*. Waite Group Press, 1993, ISBN 1-878739-39-5.

Standage, Tom. *The Victorian Internet*. New York: Walker and Company, 1998.

Stephenson, Neal. *In the Beginning Was the Command Line*. New York: Avon, 1999.

Stern, Thomas E. and Krishna Bala. *Multiwavelength Optical Networks: A Layered Approach*. Reading, MA: Addison-Wesley Longman, 1999.

Stoll, Cliff. *The Cuckoo's Egg: Tracking a Spy Through the Maze of Computer Espionage*. New York: Simon and Schuster, 1989.

Tapscott, Donald. *Growing Up Digital: The Rise of the Net Generation*. New York: McGraw-Hill, 1998.

Tapscott, Donald, Alex Lowy, and David Ticoll. *Blueprint for the Digital Economy*. New York: McGraw-Hill, 1998.

Thanlmann, Daniel. *Virtual Worlds and Multimedia*. New York: John Wiley, 1993.

Wexelblat, Alan (Editor). *Virtual Reality: Applications and Explorations*. Academic Publishers, 1993. ISBN 0-12-745045-9.

Woolley, Benjamin Blackwell. *Virtual Worlds: A Journey in Hype and Hyperreality*. 1992.

Articles

——. "A Long March." *The Economist*, July 14, 2001.

——. "Bluetooth Hype Exceeds Reality." *Network Magazine*, August, 2001.

——. *Cable Industry Statistics*. www.ncta.com/industry_overview.

——. "Don't turn Your Back on WLL." *Communications News*, May 2001.

——. "Integrated Data Platform for VoIP." *America's Network*, September 15, 1999, page 78.

——. "Internet Call Waiting Gets User-Friendly Upgrades." *America's Network*, April 15, 1999, page 14.

——. "IP/ATM Platform Tackles Brain as well as Brawn." *America's Network*, February 15, 1999, page 14.

——. "Is Sun/AOL Alliance a Mixed Blessing?" *America's Network*, May 1, 1999, page 13.

——. "Keeping the Customer Satisfied." *The Economist*, July 14, 2001.

——. "Mindspring Loses its Bounce." *TECHCapital*, September/October 1999, page 15.

——. "Multiprotocol Label Switching (MPLS)." *ATG's Communications & Networking Technology Guide Series*, sponsored by ennovate. Available at www.techguide.com.

——. "Put Some Backbone Into Your Decisions." *Communications News*, April 1999, page 22.

———. "Putting It in its Place." *The Economist*, August 11, 2001.

———. "Real-Time IP Billing." *America's Network*, September 15, 1999, page 79.

———. "Spoilt for Choice." *The Economist*, July 7, 2001.

———. *The CIO Wireless Resource Book*. A white paper from Synchrologic, www.synchrologic.com.

———. "The Internet's New Borders." *The Economist*, August 11, 2001.

———. "The Joy of Text." *The Economist*, September 15, 2001.

———. "The North American Fiber Highway System." *America's Network*, February 15, 1999, page 28.

———. "U.S. Taking the Slow Road to 3G." *ZDNet*, October 1, 2001.

———. "Unleash the Power: Building Multi-Service IP Networks With ATM Cores." A white paper published by the ATM Forum. Available at www.atmforum.com.

———. *V.92 FAQ*. www.v92.com.

3Com Corporation. "In Phase With the Future." *Net Age*, Q3 1999, page 5.

Allen, Doug. "Will Gigabit Ethernet WAN Services Make us Forget About SONET?" *Network Magazine*, July 2001.

Alster, Norm. "The New AT&T: Now for the Hard Part." *Upside*, September 1999, page 120.

Anderson, Howard. "Fixed Wireless—or Fixed Stupidity?" *Network World*, June 4, 2001.

The ATM Forum. "Unleash the Power: Building Multi-Service IP Networks with ATM Cores." Available from the ATM Forum web site, www.atmforum.com.

Bachinsky, Shmuel. "Bridging the Gap with MGCP." *Communications Systems Design*, September 1999, page 9.

Barrett, Randy, and Carol Wilson. "Trouble in DSL's Broadband Paradise." *Interactive Week*, October 25, 1999.

Blake, Pat. "Will Carriers Bypass SONET and Take Data to DWDM?" *America's Network*, special supplement.

Borrus, Amy. "Why MCI's Brat Pack Is All Over the Beltway." *Business Week*, August 30, 1999, page 172.

Bosch Telecommunications. "LMDS—Architectural Overview." From the corporate web site, www.boschtelecominc.com/lmds/architec.htm.

Botting, Chris. "Switching Software Boosts IP Telephony." *Network World*, November 23, 1998, page 37.

Bradner, Scott. "IP Phone or Internet Phone?" *Network World*, October 4, 1999, page 56.

Brayton, Colin. "The Learning Curve." *Internet World*, October 2001.

Breidenbach, Susan. "Got the Urge to Converge?" *Network World*, September 27, 1999, page 51.

———. "VoIP Variables." *Network World*, October 15, 2001.

Brewer, John. "The IP App that's Ready for Primetime." *CTI*, May 1999, page 88.

Briere, Daniel and Christine Heckart. "What Organized Crime and Convergence Have in Common." *Network World*, August 1999.

Brisard, Al. "The VoIP-PBX Connection." *Communications News*, May, 2001.

Broderson, Mikkel and Bengt Beyer-Ebbesen. "ATM Flow Control: Recipe for Demanding Apps." *Network World*, August 31, 1998, page 29.

Brothers, Art. "The Sins of IP Telephony." *America's Network*, December 1, 1998, page 54.

Brumfield, Randy. "What it Takes to Join the Carrier Class." *Internet Telephony*, May 1999, page 80.

Burrows, Peter. "A Bright Ray of Sun at AOL." *Business Week*, September 27, 1999, page 58.

Carlson, Randy and Martyn Roetter. "Justifying the Need for Fiber Networks." *America's Network*, February 15, 1999, page 28.

Carr, Jim. "Lesson 155: Service Level Agreements." *Network Magazine*, June 15, 2001.

Caruso, Jeff. "Swedish Trio Touts ATM Alternative." *Network World*, January 18, 1999.

Chappell, Laura. "Migrating to IP." *Network World*, October 18, 1999, page 63.

Chidi, Jr., George. "Satellite Internet: Wireless Medium Looks for a Niche." *Network World*, June 25, 2001.

Christopher, Abby. "Now, Play Nice." *Upside*, October 1999, page 100.

Clark, David S. "High-Speed Data Races Home." *Scientific American*, October 1999, page 92.

Clark, Elizabeth. "Lesson 157: The Resource Reservation Protocol."
Network Magazine, August 3, 2001.

———. "Lesson 158: Differentiated Services." *Network Magazine*,
September 5, 2001.

Clarke, Arthur C. "Extra-Terrestrial-Relays: Can Rocket Stations Give
World-Wide Radio Coverage?" *Wireless World*, October 1945, pages 305
to 307.

Collins, Greg and Tam Dell'Oro. "Gigabit Ethernet—The Market Takes
Off." *Business Communications Review*, April 1999, page 32.

Connor, Chris. "The Network Service-Provider Challenge." *Communications News*, September 1999, page 90.

Conry-Murray, Andrew. "Lesson 159: Internet Protocol Version 6." *Network
Magazine*, October 5, 2001.

Cookish, Mike. "Bringing Policy to LANs and WANs Via 802.1p and Diff-
Serv." *Communications News*, September 1999, page 56.

Coursey, David. "Data, Data Everywhere." *Upside*, September 1999, page
148.

Cox, John. "How to Pick a Java Application Server." *Network World*, September 13, 1999, page 64.

———. "Rethinking Wireless LANs." *Network World*, October 22, 2001.

Cray, Andrew. "Voice Over IP: Hear's How." *Data Communications*, April
1998, page 44.

Crockett, Roger O. "Why Motorola Should Hang Up on Iridium." *Business
Week*, August 30, 1999, page 46.

Cross, Kim. "B-to-B, By the Numbers." *Business 2.0*, September 1999, page
109.

Davis, Brad. "The Standards Shuffle." *Telephony*, August 23, 1999, page
52.

Delaney, John. "ASPs Will Drive Next-Generation Telcos." *ZDNet*, October
28, 2001.

Dertouzos, Michael. "The Future of Computing." *Scientific American*,
August, 1999.

DeVeaux, Paul. "Fighting Among the Switch Vendors." *America's Network*,
August 1, 1999, page 26.

———. "The New World of IP Billing." A supplement to *America's
Network*, page S6.

———. "Nortel's Anytime, Anywhere Internet Access." *America's Network*, July 15, 1999, page 10.

———. "Programming (In)flexibility." *America's Network*, September 1, 1999, page 50.

———. "Voice Over IP: Promise or Problems?" *America's Network*, May 1, 1999, page 61.

Doolan, Paul. "QoS." *CTI*, May 1999, page 121.

Dornan, Andy. "America on the Couch." *Network Magazine*, September 2001.

———. "Is There an Afterlife for ATM?" *Network Magazine*, May, 2001.

———. "Network Management in a Crisis." *Network Magazine*, July 5, 2001.

———. "Three Standards for 3G?" *Network Magazine*, September 2001.

———. "WAP Reaches the Second Generation." *Network Magazine*, September 2001.

Drucker, Peter F. "Beyond the Information Revolution." *The Atlantic Monthly*, October 1999, page 47.

Dubie, Denise. "Management Service Providers Feast on Challenge." *Network World*, June 4, 2001.

Duffy, Jim. "3Com Captain Remains Calm Despite Stormy Forecasts." *Network World*, July 5, 1999.

———. "Cisco Bringing Convergence to Small Offices." *Network World*, August 9, 1999.

———. "Cisco Buys Voice Switch Maker Summa Four." *Network World*, August 3, 1998.

———. "Cisco Unveils Packet Telephony Gear." *Network World*, August 2, 1999.

———. "Emerging Standard to Speed Up Ethernet Reconfigs." *Network World*, November 1, 1999.

———. "Lucent's Metro Re-Entry May Re-Energize Giant." *Network World*, June 18, 2001.

———. "Why MPLS Matters to Enterprise Networks." *Network World*, June 21, 1999.

Durlach, N. I., W. A. Aviles, R. W. Pew, et. al. "Virtual Environment Technology for Training (VETT)." *BBN Report No. 7661*. Cambridge, MA: Bolt Beranek and Newman, Inc., March 1992.

Dyrness, Christina. "Sayonara, Cisco." *TECHCapital*, September/October 1999, page 38.

Dzubeck, Frank. "Application Service Providers: An Old Idea Made New." *Network World*, August 23, 1999, page 49.

Edwards, Morris. "IP Telephony Ready to Explode Into Corporate World." *Communications News*, May 2001.

Ellerin, Susan. "Network Management Platforms Make the Grade." *Network World*, September 13, 1999, page 73.

Elstrom, Peter and Linda Himelstein. "What will AT&T do with Excite@Home?" *Business Week*, August 30, 1999, page 44.

Federal Communications Commission. "FCC Action to Accelerate Availability of Advanced Telecommunications Services for Residential and Small Business Consumers." *Common Carrier Action*, November 18, 1999.

Figueredo, Ken and Brian Toll. "Are You Ready for Convergence?" *Network World*, September 13, 1999, page 99.

Fisher, Dave. "Defrauding the Fraudsters." *America's Network*, May 1, 1999, page 53.

Fontana, John. "Microsoft Shapes Its Collaboration Platform." *Network World*, September 27, 1999, page 16.

Foster, Paula. "Meeting in the Cloud: Next-Gen Networks, New Enhanced Services." *CTI*, June 1999, page 82.

Fox, Loren. "Another Face for Venture Capitalism?" *Upside*, October 1999, page 127.

Freedman, Rick. "Has the ASP Market Gone Vertical?" *Tech Republic*, August 20, 2001.

Fried, John. "Cashing in, Moving on." *Washington Business Forward*, October 1999, page 24.

Friedrichs, Terra. "Talk Is Cheaper with Voice Over IP." *Data Communications*, November 21, 1998, page 21.

Galitzine, Greg. "The Wireless Last Mile." *Internet Telephony*, June 1999, page 60.

Garner, Rochelle. "Supply-Chain Leader Ventures into Internet-Based ERP." *Upside*, September 1999, page 53.

Gately, Joe. "Road-Warrioring Made Easy." *Communications News*, September 1999, page 22.

Gibbs, Andy. "MSPs: The Benefits." *Network World*, August 27, 2001.

Globalstar Corporation. *About Globalstar*. From the corporate web site, www.globalstar.com/en/about/index.html.

Gold, Steve. "Wireless Gaming Set to Soar." *Newsbytes*, August 8, 2001.

Granstrom, Peter, Dennis Niermann, and Kenneth Yong. "Building Bridges with SS7 Technology." *Communications News*, September, 2001.

Greene, Tim. "A New Twist on DSL: Voice Services." *Network World*, August 16, 1999, page 36.

———. "AT&T, BT to Craft International Venture." *Network World*, August 3, 1999, page 23.

———. "HDSL2 Could Mean Cheaper T-1s for You." *Network World*, October 4, 1999, page 6.

———. "QWEST and IXC Join Big Boys With DSL Offerings." *Network World*, August 9, 1999, page 8.

———. "Splitterless DSL Promises Faster Service Delivery." *Network World*, November 1, 1999, page 12.

———. "The Vaunted VPN." *Network World*, September 27, 1999, page 65.

———. "VPNs: IP Adds a New Twist." *Network World*, September 24, 2001.

Greenfield, David. "North American Carrier Survey: Tough Enough?" *Network Magazine*, August 2001.

Gross, Neil. "21 Ideas for the 21st Century." *Business Week*, August 30, 1999, page 132.

Hammond, Eric. "Managing the Data Mountain." Special Report to *Info World*; October 11, 1999, page S1.

Haramaty, Lior. "VoIP: The Opportunity. Why You Should Become an ITSP." *Internet Telephony*, July 1999, page 32.

Hecht, Howard. "At What Price Ubiquity?" *TechWeb*, www.teledotcom.com/analysts_alley/analyst.html. July 22, 1999.

Helyar, John. "The One Stock You Can't Ignore." *Money*, October 1999, page 106.

Hesseldahl, Arik. "Megahertz Marketing Mayhem." *Forbes.com*, August 8, 2001.

———. "Semiconductor Recovery Still on Track." Forbes.com, October 24, 2001.

———. "Silicon Confusion." Forbes.com, October 10, 2001.

Heywood, Peter. "Doing in ATM?" *Data Communications*, April 21, 1999, page 67.

Hindin, Eric. "Say What?" *Network World*, August 17, 1999, page 37.

Hollman, Lee. "The Era of IP Telephony Is Upon Us (In Stages)." *Call Center Magazine*, September, 2001.

Hotch, Ripley. "Check Mate." *Communications News*, April 1999, page 10.

Hurwicz, Michael. "Groove Networks: Think Globally, Store Locally." *Network Magazine*, May, 2001.

Isenberg, David. "Mother of All Disruptions." *America's Network*, July 15, 1999, page 12.

Jones, Steve. "The Path to 3G." *Network World*, June 25, 2001.

Judd, James. "High Wireless Act." *Upside*, March 2001.

Kazi, Shaheen. "SHDSL: DSL in Overdrive." *Communications News*, October, 2001.

Keegan, Paul. "Can Bob Pittman, of MTV Fame, Make AOL Rock?" *Upside*, November 1999, page 98.

———. "Is This the Death of Packaged Software?" *Upside*, October 1999, page 139.

Keeton, K. and R.H. Katz, "The Evaluation of Video Layout Strategies on a High-Bandwidth File Server," Proceedings of the 4th International Workshop on Network and Operating System Support for Digital Audio and Video, Lancaster, England, November 1993.

Kelly, Sean. "Is the Clock Ticking for Frame Relay?" *Communications News*, September 2001.

Kewney, Guy. "It's Time to Wake Up Wireless." *ZDNET*, October 22, 2001.

King, Rachel. "The New Breed of Telecom Companies." *TECHCapital*, September/October 1999, page 47.

Klaiber, Doug. "A New Switching Architecture for a New Competitive Environment." *CTI*, June 1999, page 78.

Korzeniowski, Paul. "Local-Loop Products Are Good for the Long Haul." *Internet Week*, September 13, 1999, page 48.

———. "Sun Setting on FDDI." *Business Communications Review*, April 1999, page 47.

Kovac, Ron. "VPN Basics." *Communications News*, April 1999, page 14.

Krapf, Eric. "Visions of the New Public Network." *Business Communications Review*, May 1999, page 20.

Kraskey, Tim and Jim McEachern. "Next-Generation Voice Services." *Network Magazine*, June 1999.

Langdon, Greg. "Voice Over DSL Sounds Promising." *Network World*, August 2, 1999, page 31.

Lawrence, Jeff. "Integrating IP and SS7 Technologies." *CTI*, April 1999, page 58.

Layland, Robin. "QoS: Moving Beyond the Marketing Hype." *Data Communications*, April 21, 1999, page 17.

Levine, Shira. "The ABCs of ERP." *America's Network*, September 1, 1999, page 54.

————. "The Generation Gap." *America's Network*, August 1, 1999, page 52.

————. "TMN: Dead or Alive?" *America's Network*, July 15, 1999, page 40.

Lewis, Jamie. "As E-Business Evolves, Boundaries Will Give Way to Virtual Enterprise Networks." *Internet Week*, September 13, 1999, page 25.

Lieberman, David. "Bridging the Digital Divide." *USA Today*, October 11, 1999, page 3B.

Liebmann, Lenny. "Bandwidth: Raw Material for the New Economy." *Communications News*, April 1999, page 84.

Liebrecht, Don. "A TCP/IP-Based Network for the Automotive Industry." *America's Network*, May 1, 1999, page 67.

Lindstrom, Ann and Richard Karpinski. "Telecom History 101." *Telephony Magazine*, June 17, 1991.

Lindstrom, Annie. "Fiber 'Firsts' Shine Bright." *America's Network*, July 15, 1999, page 26.

————. "SONET Celebrates its 15th Anniversary." *America's Network*, September 1, 1999, page 32.

————. "The Fiber Webmasters." *America's Network*, December 1, 1998, page 19.

————. "The Opti-mologists." *America's Network*, September 1, 1999, page 20.

————. "When Will Prepay in the USA Have Its Day?" A supplement to *America's Network*, page S9.

Lippis, Nick. "Enterprise VoIP: Two to Go." *Data Communications*, July 1999, page 19.

———. "Lucent vs. Cisco: The Coming Showdown." *Data Communications*, September 1998, page 19.

Lucent Technologies. "Lucent Technologies Launches Breakthrough DSL Platform to Deliver High-Quality Voice, Data and Video Services." Company press release. September 7, 1999.

Lynch, Graham. "Dropping EDGE Could Retain Edge for AT&T." *America's Network*, February 1, 2001.

———. "How GSM Beat CDMA." *America's Network*, June 1, 2001.

MacLeod, Alan. "Protecting the 'Private' in VPN." *Network World*, September 27, 1999, page 33.

Makris, Joanna. "One-Pipe-Fits-All ATM." *Data Communications*, July 1998, page 26.

———. "Ticket to Hide?" *Data Communications*, September 1999, page 49.

Martin, Michael. "Fixed Wireless No Wipeout, Despite Recent Troubles." *Network World*, June 4, 2001.

Mason, Charles. "Competition from Above?" *America's Network*, July 15, 1999, page 16.

———. "Keeping on Track with 3G." *America's Network*, September 15, 1999, page 61.

———. "The Long and the Short of 3G." *America's Network*, May 1, 1999, page 38.

Matsumoto, Craig. "Optical Nets Could Threaten IP's Future." *Data Communications*, August 20, 1999.

Mayer, Kevin and Dan Callahan. "This Old Enterprise." *Communications Solutions*, September 2001.

McLaughlin, Michael. "Optical Signaling and Control: Opening the Door to New Ways of Building Networks. Converge!" *Network Digest*, October 22, 2001.

Mejia, Robin. "MSPs: This Year's Model?" *Network Magazine*, October 2001.

Messmer, Ellen. "Financial Firms Investing in Web-Based Customer Service." *Network World*, October 11, 1999, page 43.

Michael, Bill. "SIP Ascendant." *Convergence Magazine*, June 2001.

Moeller, Michael. "Maybe This is One Race Microsoft Can't Win." *Business Week*, March 22, 1999, page 37.

Moozakis, Chuck. "High-Speed Token Ring Moves Closer to Reality." *Internet Week*, March 25, 1998.

Moran, Rich and Steve Cortez. "Evolving Toward the All-Optical Network." *America's Network*, September 1, 1999, page 39.

Mordock, Geoff. "Wireless Optical Brings a Broadband Solution." *Communications News*, May 2001.

Mosquera, Mary. "DSL High-Speed Rollout to Snowball in 2000." *TechWeb*, September 22, 1999.

"Driving IP." *America's Network*, September 15, 1999, page 18.

Musthaler, Linda. "ASPs Make Strong Case for Renting vs. Buying Apps." *Network World*, September 27, 1999, page 37.

National Academy of Sciences, National Research Council, Committee on Virtual Reality Research and Development, Computer Generation Technology Group. "Report on the State-of-the-Art in Computer Technology for the Generation of Virtual Environment." 1993.

Neighly, Patrick. "Eyes on the Skies." *America's Network*, May 15, 2001.

Newman, David. "VoIP Gateways: Voicing Doubts?" *Data Communications*, September 1999, page 71.

Noll, Michael. "Internet Tele-Phoney." *Telecommunications*, November 1998, page 42.

Nolle, Tom. "Qwest and USWest Have Their Roles Reversed." *Network World*, July 19, 1999, page 51.

————. "VoIP: Stalled at the Demarc." *BCR*, April 1999, page 12.

Olicom Corporation. "Olicom Sells Token Ring Business to Madge Networks." August 31, 1999. Company press release.

Ozur, Mark. "IP: Redefining the Telecommunications Industry." *Internet Telephony*, September 1998, page 84.

Pack, Charlie and James Gordon. "Engineering the PSTN." *America's Network*, September 15, 1999, page 16.

Panditi, Surya. "The Power of Light." *Communications News*, July 1999, page 24.

Pappalardo, Denise. "ASP Attack." *Network World*, September 27, 1999.

————. "AT&T WorldNet Still Mending Dial-up Net." *Network World*, July 5, 1999.

————. "Cisco, Motorola Buy into Wireless Broadband." *Network World*, June 14, 1999.

————. "Enron Building Bandwidth the IP Way." *Network World*, August 2, 1999.

————. "Global Crossing Rolls out VPN Services." *Network World*, October 22, 2001.

————. "Voice Over IP Still Has Hurdles to Clear." *Network World*, September 20, 1999.

Pearce, Alan. "An Unexpected Turn for 3G." *America's Network*, May 15, 1999, page 58.

Petrosky, Mary. "Policy Capabilities Help Drive RSVP's Renaissance." *Network World*, July 5, 1999, page 33.

Pfeiffer, Marc. "Capitalizing on IP VPNs." *America's Network*, May 15, 1999, page 40.

Pierson, Rick. "SS7—Stepping Stone to the Future." *Internet Telephony*, September 1998, page 70.

Port, Otis. "Machines Will be Smarter Than We Are." *Business Week*, August 30, 1999, page 117.

PriceWaterhouseCoopers' Editors of *Technology Forecast 1999*. "Internet Backbone and Service Providers: A Market Overview." *America's Network*, April 15, 1999, page 18.

Roberts, Lawrence. "Judgment Call—Quality IP." *Data Communications*, August 1999, page 64.

Robins, Marc. "IP Breathes New Life into the Traditional PBX Market." *Internet Telephony*, July 1999, page 22.

————. "Moving Beyond Plain Old VoIP." *Internet Telephony*, May 1999, page 22.

Rogoski, Richard R. "Serving Up Storage Solutions." *Communications News*, August, 2001.

Rohde, David. "All-in-One-Access." *Network World*, September 27, 1999, page 77.

————. "AT&T Wins Cable Battle, but War with RBOCs Continues." *Network World*, August 2, 1999, page 12.

————. "Is Qwest Losing Its IP Religion?" *Network World*, June 28, 1999, page 38.

Romascanu, Dan. "Standard Puts Keen Eye on Switched Nets." *Network World*, October 18, 1999, page 51.

Rosen, Evan. "The Great Global 3G Challenge." *Network World*, August 6, 2001.

Rowe, L.A., K. Patel, B.C. Smith, and K. Liu, "MPEG Video in Software: Representation, Transmission and Playback," Proceedings IS&T/Society of Photo-Optical Instrumentation Engineers 1994 Int. Symp. Electronic Imaging: Science and Technology, San Jose, CA, February 1994.

Rybczynski, Tony. "Follow the Leaders." *Communications News*, April 1999, page 26.

Rysavy, Peter. "Broadband Wireless: Now Playing in Select Locations." *Data Communications*, October 1999, page 73.

Sawhney, Mohanbir and Steven Kaplan. "Let's Get Vertical." *Business 2.0*, September 1999, page 85.

Scanlon, Bill. "Speeding Along the MPLS Byway." *Interactive Week*, July 24, 2001.

Schlender, Brent. "The *Real* Road Ahead." *Fortune*, October 25, 1999, page 138.

Schwartz, Ephraim. "'Parasitic Grid' Wireless Movement May Threaten Telecom Profits." *Infoworld*, August 24, 2001.

Skran, Dale. "H.323: Forward, March!" *CTI*, July 1999, page 102.

Slater, Bill. "Have it Your Way!" *Communications News*, August 1999, page 74.

Sloan, Allan. "How We'd Make Our Millions." *Newsweek*, June 21, 1999, page 55.

Smith, B.C. "Implementation Techniques for Continuous Media Systems and Applications," Ph.D. dissertation, U.C. Berkeley, June 1994.

Smith, B.C. and L.A. Rowe, "Algorithms for Manipulating Compressed Images," *Computer Graphics and Applications*, Vol. 13, No. 5, September 1993, pp. 34–42.

Songini, Marc. "IBM Net Gear Set to Go to Cisco." *Network World*, September 6, 1999, page 1.

———. "Packeteer Brings SNA reliability to IP." *Network World*, September 6, 1999, page 21.

Southwick, Karen. "Java Takes Off." Excerpt from *High Noon: The Inside Story of Scott McNealy and the Rise of Sun Microsystems. Upside*, October 1999, page 153.

Spears, Mit. "The Empire Strikes Back." *Upside*, September 1999, page 150.

Steinert-Threlkeld, Tom. "The Net: Not Just Data Anymore." *Internet 2002*, August 31, 1998.

Stenson, Tom. "The QoS Quagmire." *Network World*, September 6, 1999, page 53.

Stephenson, Ashley. "Controlling Oversubscription." *Communications News*, May 1999, page 60.

Sullivan, Ann. "Integration Drives PeopleSoft's Retooled CRM Apps." *Network World*, June 4, 2001.

Sweeney, Dan. "Cracking the Code." *America's Network*, October 1, 2001.

———. "The Interminable Last Mile." *America's Network*, May 1, 2001.

Takahashi, Dean. "Start-Up Extends Reach of DSL Lines." *Wall Street Journal*, October 14, 1999, page B8.

Tanner, John C. "ATM Lives!" *America's Network*, May 1, 2001.

Taylor, Kieran. "Leveraging SS7 for Converged Voice and Data." *CTI*, September 1998, page 58.

Tehrani, Rich. "Nortel Weighs in on Internet Telephony." *Internet Telephony*, July 1999, page 6.

Teledesic Corporation. "Teledesic Fast Facts." From the corporate web site, www.teledesic.com.

Thomas, R. Todd. "SAN Sees Through Data Storage Drain." *Tech Republic*, September 27, 2001.

Titch, Steven. "Mediation Devices: Shouldering the Load." *Data Communications*, page 77.

Tolly, Kevin. "Converged Telephony in 1999?" *Network World*, May 31, 1999, page 24.

———. "One Giant Step for High-Speed Token Ring." *Network World*, June 1, 1998.

———. "When Usage Matters: Broadwing's Gigabit in the MAN." *Network World*, August 6, 2001.

Tracey, Lenore V. "Internet Telephony: Who's Buying?" *Telecommunications*, September 1999, page 31.

Turner, Brough. "The Impact of CTI on Wireless Communications." *CTI*, May 1999, page 44.

United States Congress. Virtual Reality: Hearing Before the Subcommittee on Science, Technology, and Space of the Committee on Commerce, Science, and Transportation, and United States Senate. New Developments in Computer Technology. Washington, DC: U.S. GPO. (United States. Congress. Senate Hearing, 102 to 553). LC CALL NUMBER: KF26.C697 1991e. 1992.

Van Beurden, Lisa. "Metro Networks: It's Time for an Overhaul." *FiberSystems International*, June 2001.

VanderBrug, Gordon. "Making QoS Work with Voice Over IP." *Communications News*, April 1999, page 56.

Vaughan-Nichols, Steven J. "Instant Messaging—Better Safe Than Sorry." *ZDNet*, October 2, 2001.

———. "Ultrawideband Wants to Rule Wireless Networking." *ZDNet*, October 30, 2001.

Vickers, Lauri. "Ethernet: The Perfect 10?" *Network Magazine*, June 2001.

Wall, Kendra. "Splitting the Spectrum." *Upside*, September 2001.

Wallman, Roger. "The Case for H.323 In Video Conferencing." *Internet Telephony*, June 1999, page 78.

Wallner, Paul. "The Time is Right." *Communications News*, July 1999, page 54.

Wexler, Joanie. "QoS: What Can Service Providers Deliver?" *Business Communications Review*, April 1999, page 25.

Willey, Richard. "How Cryptic!" *Communications News*, April 1999, page 18.

Williams, Molly. "Chip Makers to Demonstrate InfiniBand, Standard Aimed at Breaking Data Logjam." *Wall Street Journal*, August 23, 2001.

Willis, David. "The Year of the ATM WAN?" Network Computing, July 26, 1999.

Zimmerman, Christine. "ATM! The Technology That Would Not Die!" *Data Communications*, April 21, 1999, page 46.

Web Resources

http://mpeg.telecomitalialab.com/standards/mpeg-4/mpeg-4.htm. MPEG-4 overview.

http://mpeg.telecomitalialab.com/standards/mpeg-7/mpeg-7.htm#a_Toc533998968. MPEG-7 overview.

http://philotfarnsworth.com/. The Philo T. Farnsworth web site.

http://www.dcs.ed.ac.uk/home/mxr/gfx/2d/BMP.txt. Bitmap overview.

http://www.dcs.ed.ac.uk/home/mxr/gfx/2d/GIF87a.txt. GIF overview.

http://www.dcs.ed.ac.uk/home/mxr/gfx/2d/JPEG.txt. JPEG overview.

http://www.dcs.ed.ac.uk/home/mxr/gfx/2d/TIFF-5.txt. TIFF overview.

http://www.dvddemystified.com/dvdfaq.html#a3. Frequently asked questions about DVD.

http://www.faqs.org/faqs/compression-faq/part1/. Frequently asked questions about digital compression, Part One.

http://www.faqs.org/faqs/compression-faq/part2/. Frequently asked questions about digital compression, Part Two.

http://www.faqs.org/faqs/compression-faq/part3/. Frequently asked questions about digital compression, Part Three.

http://www.nwfusion.com/primers/ipm/ipmscript.html. Multicast audio primer.

Photo Credits

Figure 1-8 used courtesy of the Sarnoff Institute.

Figures 1-9 and 1-10 used courtesy of the Philo T. Farnsworth Archives.

Figure 1-11 used courtesy of the Illinois Institute of Technology.

Figure 1-2 and 3-9 used courtesy of Proximity and Tandberg USA.

APPENDIX E

Closing Thoughts: The Medium Is the
Message—Or Is It?

My background is a bit professionally schizophrenic. I grew up in Spain and later studied Romance Philology at UC Berkeley (that's the study of the origins of Romance Languages, for you engineering types), but took a lot of Marine Biology courses as well. When I graduated, I went in search of the perfect career through which I could use my knowledge of the Spanish language and its roots, together with my love of all things water related. I certified as a SCUBA diving instructor, opened a shop with a partner, and started an underwater cinematography and commercial diving business. In 1980, we purchased a business in Saint Maarten in the Netherlands Antilles to use as a base of operations for the work we were doing in that part of the world. The business as a whole did well for quite a few years— from 1979 until 1981, actually—until my partner and I decided to go our separate ways. So I folded the corporation, resigned as president, and went off in search of a job. I sort of had to hurry because I was scheduled to marry Sabine not long thereafter, and it just wouldn't do to be unemployed while getting married.

Dive shops are pretty cool places to hang out in, and they attract their fair share of groupies. One of the people who was a regular at our dive shop was a fellow named Jack Garrett. Jack was in the first SCUBA class I taught at Sea Hut, and he was a natural diver. He was so good that I certified him as an assistant instructor not long after he got his basic SCUBA certification so that he could help out with classes.

Shortly after the business closed, Jack came to me with a proposition. He was a district manager with Pacific Telephone, and Pacific was embarking on a radical new technical training program. They were looking for people with management experience, but no technical background to come to work for the company and be trained as network analysts. Please understand: I didn't know a network from a box of spaghetti. Was I interested, he asked? "Does it pay a regular paycheck," I countered? "Yes it does," he replied, and I remarked that this sounded like a terrific career choice. The next thing I knew, with Jack's letter of recommendation in hand, I had purchased a suit,[1] had been through the BSQT6,[2] had sweated through the initial interviews and physical,[3] and the next thing I knew I was employed at the *Network Operations Control Center* (NOCC) in San Francisco. I was both elated and terrified because my experience with computers was somewhat limited[4], and they made it quite clear that it would mean the end of civilization as we knew it if we were to do something stupid.

By now, you're probably wondering what all this has to do with media selection, so let me tell you. I have been professionally involved with education for 20 years, ranging from a high school teacher to a SCUBA instructor to a hands-on technical trainer on a wide range of telecomm equipment to what I do now through my own company. The buzzwords come and go, the presentation techniques pass in and out of favor, and the argument over teaching style versus learning style waxes and wanes in popularity like the phases of the moon. One constant, however, is the need for learning effectiveness.

Corporate Trends

In today's corporation, there is a renewed focus on employee performance as a critical factor in corporate success. According to studies conducted by the *American Society of Training and Development* (ASTD), there are five key behavioral elements that affect employee and corporate performance. They are

---

[1]I owned lots of suits, but they were all made of neoprene.

[2]Bell System Qualification Test of Basic Skills. Sort of like taking the SATs all over again.

[3]Meanwhile, we *did* get married, and while I sat by feverishly waiting for the phone to ring to tell me whether I had a job, Sabine went to Hawaii for our honeymoon. The call came while she was gone.

[4]As in zip. Zero. Nada. Didn't know a computer from a water cooler.

- A focus on those results that make a company competitive
- Measurable change
- People with the right skill, knowledge, and behavior to perform as desired
- Systems and processes that connect work effort to desired results
- Methods for analyzing and closing gaps between current and desired performance

Furthermore, their studies indicate that most successful organizations today share the following characteristics:

- They tend to have flatter, more horizontal (matrixed) managerial structures.
- Work is done by teams organized around business processes.
- Highly skilled workers are empowered to act as they see fit.
- There is significant collaboration among teams, between labor and management, and with suppliers.
- There is a widely accepted and recognizable focus on quality, customers, and continuous improvement.
- They employ flexible technologies.
- They have implemented a formal change management process.

The ASTD has also identified 10 trends that are underway in most corporations and that affect the degree to which training is required:

- Skill requirements on the part of employees will continue to increase in response to rapid technological change.
- The American workforce will be significantly more educated and more diverse.
- Corporate restructuring will continue to reshape the business environment.
- Corporate training departments will change dramatically in size and composition.
- Advances in technology will revolutionize the way training is delivered.
- Training departments will find new ways to deliver services.

■ Training professionals will focus more on interventions in performance improvement.

■ Integrated high-performance work systems will proliferate.

■ Companies will transform into learning organizations.

■ Organizational emphasis on human performance management will accelerate.

Clearly, employees play a central role in the success of the modern corporation, but there is a caveat: They must be trained to understand their role in the organization and to carry out that role in the most effective manner possible. The need for speed dictates that training methods must be timely and accurate—smart bomb education as opposed to the shotgun approach.

Media History in Adult Education

Multimedia education, believe it or not, has been around for about 60 years. Some of you will remember the days in elementary school when you fought over who got to sit behind the filmstrip projector and advance the filmstrip to the next image when the beep sounded on the scratchy record spinning at the front of the room. Not very elegant and certainly not integrated, but multimedia nonetheless. That technique and others like it evolved into truly integrated solutions encompassing CD-ROMs, videotape programs, audio lessons, computer-based and computer-assisted training packages, virtual reality *Multiuser, Object-Oriented Environments* (MOOs) and *Multiuser Dungeons* (MUDs), instructor-led sessions, and various combinations of all of these.

Years ago, Marshall McLuhan observed that "the medium is the message." His words were taken to heart by a generation of educators and trainers who decided that the presentation medium is the key to any successful training program. To a point, that's true; unfortunately, a radical splinter group decided that if one medium is good and causes students to learn, then *lots* of media must be *great* and will cause students to *really* learn. Thus was born today's fascination with multimedia as the solution to the world's educational challenges. Although there is nothing wrong with multimedia or with any of the media that comprise it, educators must be judicious in their use of alternative media to ensure that they maintain presentation balance in the classroom. Students—especially adult students —act like the variable rate diffusion filters we used to use in chemistry classes. They absorb material at a good clip for a period of time, but they do hit limits, after which *nothing* will be absorbed. It is this variable diffusion problem that alternative media are particularly good for.

McLuhan argued that modern electronic media, such as computers, radio, television, and films, have far-reaching sociological, aesthetic, and philosophical consequences, to the point that they actually alter the ways in which we experience the world. His observation is that "the medium is the message," but I prefer to modify that phrase (if you'll pardon the pun) by observing that "the medium is the *massage.*" That is, the medium should certainly *shape* the message and provide emphasis and increase penetration and absorption. It should not, however, be used just for the sake of injecting an alternative medium into a learning environment because one hasn't been used in a while, or because there might be a perception that the teaching organization isn't with it because they haven't converted their entire curriculum to laserdisc.

One implication of the shift towards technology-based training or computer-mediated training as it is often called, is the need for training professionals to work closely with technologists to develop delivery systems. Training departments must partner with information technology and information systems organizations. In a significant number of corporations, those departments are leading the way in the development of expert systems and electronic performance support systems and in research on the delivery of training via computer networks.

Back to Diving

The typical basic SCUBA certification course consists of about 16 hours of classroom time, 16 to 20 hours of pool time, and a minimum of three ocean or lake dives. In the classroom, students learn about such dry-as-bone-meal topics as Archimedes' Principle, Boyle's and Charles' Laws, equipment design and maintenance, the proper use of the U.S. Navy Decompression Tables, the physics of diving (sound propagation, the loss of color at depth, overpressure accidents), diving first aid, safety and rescue, oceanography, and biology. Dry topics all, but topics that are critical for safe diving.

In the pool, students learn how to use the equipment, how to rely on a diving buddy, how to recognize the signs of fatigue or panic, and how to administer first aid or rescue in the water. They learn about physical limits, preparedness, and cooperation. Some of it seems corny, but a well-trained diver who responds immediately and correctly is a safe diver. The ocean is not a particularly friendly place for divers, and the only way to visit is on her terms—first rule of diving.

When the instructor is satisfied that divers are ready to test their skills in open water, the first dive is scheduled. It is usually a free dive (snorkels, no SCUBA) and is conducted as a means to test buoyancy control, surf entry and exit ability, rescue techniques, and general comfort in the water. A few weeks later, the first SCUBA dives are conducted. This is the acid test, during which students *really* test their skills underwater. This is where all the pool sessions pay off because it is during this first dive that Murphy arrives with a vengeance and wreaks havoc on most students. Masks flood. One fin disappears in the surf zone, never to be seen again.[5] Kelp and fishing line snag tank valves, weight belt buckles, and all the little gizmos that new divers festoon themselves with. If they've been paying attention, these annoyances are simply that—annoyances. The new divers have learned techniques for dealing with these annoyances, and they *do* deal with them. They have a good time, and at the end of the dive, are jazzed beyond belief.

It is the instructor's job to ensure that students learn what they need before embarking on their first dive and *certainly* before they are handed their certification cards. The classroom, pool, and ocean sessions are critical components to the student's preparation as a

---

[5]Somewhere off the California coast, there is an island built completely out of all the fins that my students lost over the years. The number is somewhere around 6×10^4.

certified diver, but it is what actually goes on in those sessions that makes for a truly prepared diver. In our business, we used—you guessed it—multimedia, or perhaps better stated, multiple media. Let me explain.

In the Classroom

Although a considerable amount of the material presented in the SCUBA classroom is theoretical, we always augmented it with practical material wherever possible. A certified and licensed SCUBA instructor conducted all classroom, pool, and ocean/lake sessions because the law requires it. In the classroom, the instructor would typically conduct a lecture for some portion of the three-hour session, but would augment the lecture with a 35mm slide show synchronized to an accompanying audiotape, designed for each of the lectures in the course.[6] To those slides the instructor would add his or her own slides and hands-on demonstrations that were germane to the topic at hand.

We also used hands-on exercises to illustrate key issues. For example, when we studied the effects of deep diving on the human body, I locked the students in a decompression chamber,[7] compressed them down to 150 feet, and allowed them to experience the effects of nitrogen narcosis in a dry, safe environment. I had them attempt to perform simple tasks while exposed to the narcotic effects of high-pressure nitrogen, such as calculating decompression tables for the dive. Once they were compressed to depth, giggling uncontrollably from the effects of the nitrogen, I had them blow up balloons to illustrate Boyle's Law: as we slowly "surfaced," the pressure on the outside of the balloons dropped off, but the pressure inside remained the same. By the time we reached zero depth, the balloons had all popped, proving the linear relationship between pressure and volume ($P_1V_1 = P_2V_2$): Boyle's Law. What did that exercise have to do

---

[6]This was a professionally prepared set of materials from Jeppesen Corporation.

[7]Purchased from the Navy at a surplus auction for $50—best investment we made as a business.

with real-world diving? Simple: hold your breath during the ascent, and your lungs can rupture just like the balloons.[8]

When we discussed marine biology and diving physics, we often used pool exercises to further illustrate the concepts covered. For example, in the air, we often use sound to not only get the attention of another person, but also to determine where they are through the ear's directional abilities. In water, that doesn't work because sound travels so much faster in water than in air. A diver can certainly get the attention of another diver by banging a rock on his or her tank, but the direction of the sound cannot be determined.

To illustrate this, we put the divers in the pool with SCUBA gear, wearing masks with blacked-out faceplates. We would then lower a hydrophone[9] into the pool at some random place, play music through it, and offer a substantial prize to the first student who found the source of the music. In all the years we did this, no student *ever* found the hydrophone. This exercise illustrated very effectively the fact that directional capability goes away underwater because sound travels so much faster in water than it does in air due to the different densities of the two media.

Giant kelp is a seaweed that grows in dense, thick forests along the California coast. It grows to enormous sizes and is, in fact, the fastest growing plant on earth—more than a foot per day in larger plants. It attaches to the rocks on the sea bottom, grows a long, thin rubbery stalk to within a few inches of the surface, and then sprouts four-foot leaves that spread into a thick canopy. The leaves are made buoyant by pickle-sized, gas-filled bladders,[10] and because the plants are so ubiquitous, divers often have to go through them to get to a dive site. They have two choices: they can go *under* the canopy, which requires air consumption, or they can go *over* the canopy, which consumes no air from the tank, but requires dexterity in a

---

[8]These overpressure accidents result in such trauma as cerebral air embolism, subcutaneous emphysema, mediastinal emphysema, and other nasties. They ruin your day.

[9]Basically, an underwater speaker.

[10]The bladders are filled with carbon monoxide. Each bladder contains enough CO to kill a chicken in less than a minute—no joke.

skill known as a *kelp crawl.* If done properly, a diver can crawl across the canopy as quickly as they can crawl across a hardwood floor. If done *badly,* the kelp will snag on every weight, strap, buckle, CO_2 cartridge trigger, and knife handle—at best slowing the diver down, at worst releasing buckles and sending weight belts plummeting to the bottom.

To teach the kelp crawl, we filled the pool with weight suits, which float just below the surface when they are saturated with water. With all the arms, legs, and straps, they simulate the snagging characteristics of kelp pretty realistically, which allowed us to teach the students how to successfully crawl across the canopy.

One important aspect of diving that students are required to learn is the proper use of the U.S. Navy Decompression Tables, which teach divers how many minutes they can remain at a particular depth without having to make a decompression stop.[11] This is a critical skill, although beginning divers are taught that they should *never* dive in such a way that they need the tables.[12] Typically, students are given a series of case studies that involve multiple, repetitive dives (nitrogen accumulates in body tissues, so divers have to take into account the combined effects of consecutive dives). They have to solve the problem by determining how many decompression stops they must make to off-gas following a dive to a particular depth for a certain amount of time. The process is tedious and error prone, so to help the student gain confidence I created an audio tape program that the students would play while they were solving the problems. The tape would talk them through the process, and at various times, the student would be directed to stop the tape, do part of the

---

[11]While under pressure at depth, the air breathed by the diver dissolves into his tissues. When he surfaces, the air comes *out* of solution. If divers surface too quickly, the air comes out of solution quickly, often in the form of bubbles that can lodge in tissues, joints, and backwaters of the central nervous system, resulting in extreme pain, paralysis, and death in severe cases. This phenomenon is called *decompression sickness* or more commonly, *the bends.* It is prevented either by staying within the depth/time limits stipulated in the Navy tables or by surfacing gradually, enabling the gas to come out of solution slowly.

[12]Divers are taught that the only reason to dive deep is to learn that there's no reason to dive deep. It's cold, deep, dark, and nothing lives down there.

exercise, then start the tape, at which time the narrator would explain what their results should be at that point. The tape provided a form of hand holding for the students and allowed them to have a virtual instructor at their beck and call to help them learn a particularly thorny and complex concept.

For oceanography lessons during which the instructor had to convey to the students the forces they will encounter in the ocean and the dynamic nature of coastal water (surf, rip currents, subsurface surge), we conducted lectures to discuss water movement with schematic illustrations and line drawings. Once the basic concepts had been drilled home, we used 35mm slides to illustrate the various regions of water surface activity and films that we created to show such things as a diver caught in a rip current, divers attempting to make surf entries and exits, the power of underwater surge and the danger it poses to unsuspecting divers, and proper techniques for dealing with all of these. We also used film and video to demonstrate proper boat, dock, and reef entries, rescue techniques, emergency procedures, and equipment maintenance. We would often show the films and slides on a screen set up on the pool deck, so that students could watch the demonstrations and *immediately* try the techniques in the water while they were still fresh in their minds. That also gave us the ability as instructors to iteratively train the students until the techniques were perfected.

Please note that although the principal delivery technique used to teach these would-be divers was instructor led, we relied heavily on audio, video, film, still images, hands-on practice, case studies, and iterative practice to augment the lessons presented in the traditional classroom. Note also that none of the media served as replacements for any other: The audio tapes supplemented the decompression case study exercises, the films and videos supplemented the instructor-led classroom work, the still images, illustrations, and line drawings enhanced the pool sessions, and the hands-on practice and pool work contributed to the lessons that would later be conducted in the ocean. The combination of these multiple media made for a dramatically better and more effective experience for the students, and in our minds, made them much better equipped and more confident divers.

From Diving to Data

The techniques described in the preceding section work as well in the telecommunications classroom as they do in the dive shop and pool.[13] The technical nature of much of the content is such that instructor-led education is often the best presentation technique initially because students often need one-on-one time or hand holding to grasp the initial concepts. As their confidence builds, however, other techniques can be employed. In fact, many of the media described previously are useful adjuncts in the traditional classroom and provide supplemental richness and alternative learning options for students.

In the following sections, I'd like to discuss the benefits and liabilities of some of the more popular options, including instructor led, audio tapes, videotapes, self-paced workbook exercises, *computer-based training* (CBT), and distance learning.

Instructor Led

In most traditional business environments, it can be argued that time is a constant, and knowledge is the variable. Today, however, knowledge is a desired constant while time is the variable. Most corporations have large, heterogeneous audiences to educate and inform. Management wants them all to ultimately have the same basic knowledge about the subject at hand—hence the constant knowledge concept. Unfortunately, most companies do not have the luxury of putting all employees in a room at the same time to affect the knowledge transfer. Time is clearly a variable, and the instructor-led option is a viable option.

In traditional business situations where time is constant and knowledge varies, instructor-led classes work extremely well for bringing all participants up to a required knowledge level. Instructor-mediated presentations are ideal for conveying large

---

[13]Although I wouldn't recommend teaching hands-on equipment courses in an ion-filled swimming pool. The sessions would be inspiring, but rather short.

quantities of detailed information that may require audience inter-activity for comprehension. Instructor-led classes can work well in both large (generic, easily understood material presented to a functionally, culturally, or educationally heterogeneous group) and small groups, including executive audiences where the material to be presented is highly sensitive or extremely interactive, such as material presented in a strategy session.

Instructor-led sessions are particularly effective on one hand because they allow the instructor to gauge the level of comprehension of the audience and modify the presentation in real time to affect the highest absorption rate. On the other hand, instructor-led sessions tend to be expensive and logistically difficult to coordinate because they require the ongoing presence of a living, breathing body. In the diving examples described previously, we used instructors to ensure that all participants shared a required level of competency before proceeding into the more complex, and potentially hazardous, open water phases of the course.

Videoconferencing is perhaps the alternative training medium that is most closely related to the classroom environment. It enables real-time interaction, enables the instructor to see the students and vice versa, and enables the instructor to utilize all manner of alternative media, just as he or she would in the traditional classroom.

Audio Tape

For situations that require the dissemination of broad-brush information that is high-level in nature, audio tapes are effective and low cost, provided they are well written and well produced. Audio also works well when a significant percentage of the target audience commutes because they can listen to tapes in their cars. During that time, they represent a captive audience.

Audio also works well when the material to be presented can be chunked into small pieces and used in conjunction with other presentation media. Consider the diving case studies we augmented with audio tapes: In those situations, the audio tape provided a virtual instructor who was available at the student's request to help them work through the problems. The tapes didn't replace the written exercises, nor did they replace the instructor: They augmented them. They *massaged* them. Ooh.

Audio can fall down for several reasons. If the information to be conveyed is overly complex or technical, or requires visual cues for maximum comprehension, or is overly long, audio's effectiveness falters. Alternatively, I have listened to audio presentations that were so well written and engineered that they evoked exactly the mental images that the writer wanted to create. The cost of audio is dramatically lower than video because the data collection and post-production efforts are far less complex and time consuming.

Finally, audio is a medium that can be used anywhere and by virtually anyone.

Video

Video is most effective when the message to be conveyed is conceptual or is of a nature that enables it to take advantage of the multiple sensory inputs that video provides. As a general rule, "humans tend to be multimedia devices"[14]; therefore, multiple-media interfaces work well for conveying information from device to device. In the diving classroom, we used video and film to show students techniques that we could never have described as effectively through other means. We also used the medium to transport the students to the ocean, or the lake, or a river, to help them understand what they would confront during their first dives. We showed them animals, plants, physical locations,[15] and open water techniques, always delivering a consistent, *correct* message.

---

[14]Thanks to Gary Kessler for this quote that I brazenly stole.

[15]The most popular tape had a shot of the Bathroom, a popular gathering spot in 60 feet of water near the end of the Monterey Breakwater. Years ago, someone dumped a pile of bathroom fixtures in Monterey Bay. They were collected over the years by divers and relocated to their current venue. There is a pedestal sink, two toilets, and a clawfoot bathtub with showerhead (someone even hung a shower curtain). Someone even purchased a copy of *A Fishwatcher's Guide to Pacific Coast Fishes*, printed on plastic, and left it on top of one of the toilets. The reader can imagine the photographs that have been taken there over the years.

Video is also effective for large audiences. It is the most expensive medium of all, but is also among the most effective. A well-written and well-produced video is acceptable to most everyone and can be viewed whenever time is available. When large audiences are involved, it is quite cost effective, in spite of its relatively high up-front production cost. Furthermore, it can be viewed repeatedly for maximum effectiveness and can be broadcast over closed circuit, cable, public access, corporate, or commercial television networks.

The downsides of video are worth considering. First, it is inherently linear, and thus, the message that is linked to it is dependent on the linear nature of the medium. Second, it places the viewer in a relatively passive role that becomes more passive as the amount of video increases. Programs or courses that rely on long video segments must ensure that steps are taken to reduce this passivity.[16] Whatever the case, video and audio components should be used judiciously; they reduce the degree of interactivity and can lessen the impact of the overall program.

Self-Paced Workbook Exercises

Workbooks are effective when students require asynchronous learning. Workbook material can be studied whenever the student has time for it; the issue is that this kind of learning requires a dedication on the part of the learner that isn't necessarily there for the student watching a video—or if it is there, it's often in a different form. Workbooks require that the student engage with them and actually do some proactive work. Video, on the other hand, especially *good* video, is like television. It requires virtually no interactivity and can be a marvelously seductive learning environment. Previous cautions still apply, however.

One word of caution: You can lead a diver to the ocean, but you can't make him or her dive. Giving students the ability to learn on

---

[16]In professionally produced video programs, scriptwriters use routine scene changes to engage the viewer's attention and prevent boredom. In most productions, scene changes occur every three to six seconds. Television and cinema programs exhibit similar numbers.

their own time (often called *self-directed learning*) is good, but it places a large degree of responsibility on the shoulders of the student to actually do the work. There has to be a clear incentive in front of the student to ensure that he does it. In the case of the dive student, it was easy: Learn it or potentially place yourself in harm's way. Of course, we also had written tests and practices that students had to pass prior to certification, but the subliminal incentive of potential danger was fairly motivational. In the traditional classroom, incentives tend to be less obvious and must be monitored more closely.[17] In the corporate domain, this incentive most effectively comes from a top-down mandate to complete the training. This is not always easy to bring about, however, so other incentives must often come into play.

Although workbooks aren't always the best choice for stand-alone information transfer, they are effective as accompaniments to other media. The course materials that we provided to each student comprised a textbook, a study guide and workbook, and a laminated set of the Navy Tables. Students were required to complete reading assignments and homework exercises prior to each class that were tightly connected to each other and to the materials presented in the classroom and pool during each instructor-led session.

---

[17]We had other incentives as well. When we taught advanced courses, students were required to complete six dives in two weekends: a night dive, a deep dive, and a high-altitude dive the first weekend, and a rough water dive, boat dive, and navigation/search and recovery dive on the second weekend. By the time the students prepared to do the final dive on the final Sunday, they were exhausted and quite hungry. That dive required that they follow a complex compass course without surfacing. If they did it correctly, they found me, sitting in a folding chair in front of an enormous, waterproof box in 75 feet of water. The box weighed about 150 pounds on dry land. Written on a waterproof slate on the side of the box was this message: THIS BOX CONTAINS ALL OF THE SODAS, CHAMPAGNE, BEER, WINE, AND FOOD YOU CAN POSSIBLY EAT AND DRINK. YOUR JOB IS TO GET IT INTO THE BOAT. They then had an incentive to use their newly-acquired salvage skills to carefully raise the box to the surface (too much buoyancy and it will launch itself out of the water like a Polaris missile), make it buoyant at the surface, then transfer it into the boat. They *always* got it into the boat, and it was *always* empty when we got back to the dock.

CBT

Computer-based information transfer can be quite effective. If poorly produced, however, it can also be quite ineffective. For example, if the CBT product consists of nothing more than an electronic page-turner, then it becomes nothing more than a very expensive workbook. If, however, it forces the learner to interact with the information, to wrestle with it, to actually *use* the information, then learning takes place, and the CBT works well.

We didn't have computer-based materials when I was in the diving business because PCs hadn't arrived yet. Today, however, there are many fine diving-related CBT packages, CD-ROM–based courses, interactive learning aids, online guides to marine wildlife, and decompression calculators. Many of today's certification courses rely heavily on CBT packages to teach students the basics, but instructor led remains the dominant delivery technique for core skills and for good reason. Computer technology has even replaced the traditional analog gauges that divers carry: hose-mounted consoles today include digital depth gauges, dive timers, tank pressure gauges, and decompression meters.

Interactive CBT packages that incorporate multiple media, such as text, video, and sound, are serious contenders in the multimedia CBT basket of goods. Many CBT developers combine material from a content provider with their unique *Learning Management System* (LMS) platform to produce a low-cost, reasonably good-quality product for large audiences.

Distance Learning

Distance learning is a specialized form of instructor-led presentation. It can be very effective if the broadcast medium is good; unfortunately, on lower-end systems, such as Internet-based PC systems that deliver a picture that is not up to the quality level that most people associate with television, it can be distracting. However, for large, dispersed audiences, distance learning is quite effective.

Internet and Intranet Delivery Options

Most professional training organizations today use both Internet and intranet transport to deliver significant amounts of just-in-time training to the desktop. Companies are positioning themselves to amass market share for their products, and many are already taking advantage of intranet-based training through the deployment of learning portals and the previously mentioned LMS concept.

Conclusion

All of the media discussed here have their benefits and liabilities. There is no one technique that does it all; there are too many sensory inputs out there that affect the way people learn, and environmental variables that alter the relative effectiveness of each medium, not to mention economic, social, temporal, political, and logistic variables. Each medium must be assessed in terms of the greater context of the task at hand, and the role that that medium might play *collaboratively* with other delivery options. Contrary to popular belief, instructor-led training is here to stay, but so are all the others. This is not a matter of competition: It is a matter of collaboration and mutual support, with effective learning as the ultimate outcome. In reality, the two most important aspects of this drama are the content that is delivered via a selected subset of media and the proper selection of that subset. One without the other simply doesn't work.[18]

McLuhan's belief that "the medium is the message" is off-base a bit, as far as I'm concerned. It is certainly a significant component and can dramatically affect the successful delivery and absorption of the message. The message, however, lies in the content, and the

---

[18]This matchmaking process is best facilitated by a skilled and qualified instructional designer who can guide the selection of the appropriate delivery media and match it to the content.

delivery of the content is inextricably linked to the medium. Both are critical components; neither is dispensable.

Oh, I almost forgot: diving. Check out the *National Association of Underwater Instructors* (NAUI) at www.naui.com and the *Professional Association of Diving Instructors* (PADI) at www.padi.com. Then, drop me a line, and I'll convince you.

INDEX

Symbols

256-bit color, 166
56 Kbps modem, 89
802.11
   MAC layer, 118–119
   PHY layer, 118
802.1p standard, 82
802.1Q standard, 82
80:20 Rule, 74

A

AAL (ATM Adaptation Layer), 148
access options, LANs, 69
Accord Networks, 204
acoustics, videoconferencing rooms, 229
ACT Teleconferencing, 210, 216
addresses
   ATM, 151–152
   IP, 159–161
ADSL (Asymmetrical Digital Subscriber
     Line), 97
   DSLAMs, 100
   FDM, 98
   splitters, 98
Aethra, 204
ALUs (Arithmetic-Logic Units), 58

ASTD (American Society of Training and
     Development), 312
AT&T
   backbone services, 213
   bridge services, 209
   PicturePhone, 237
ATM (Asynchronous Transfer Mode),
     144–145
   addresses, 151–152
   cell headers, 149–151
   evolution, 146
   Physical Layer, 149
   protocols, 147
   services, 152–155
ATM Forum services, 154
ATM Layer, 148
audio tape educational materials, 322
avatars, 199

B

B-Channels, ISDN BIR, 93
backbone providers, 213–214
bandwidth, SONET, 131
behavioral elements affecting
     employee/corporate performance, 313
black-and-white television, 35
Blue screen, 46

X–Z

ABOUT THE AUTHOR

Steven Shepard is a professional writer and educator specializing in international telecommunications. He is the author of four well-received books: *Telecommunications Convergence*, *SONET/SDH Demystified*, *Optical Networking Crash Course,* and *Telecom Crash Course*. Formerly with Hill Associates, now President of the Shepard Communications Group, he conducts seminars and workshops on telecom topics around the country. He lives in Williston, Vermont.